Mary Ellen Bamford

Up and Down the Brooks

Mary Ellen Bamford

Up and Down the Brooks

ISBN/EAN: 9783337140342

Printed in Europe, USA, Canada, Australia, Japan

Cover: Foto ©berggeist007 / pixelio.de

More available books at **www.hansebooks.com**

UP AND DOWN THE BROOKS

BY

MARY E. BAMFORD

BOSTON AND NEW YORK
HOUGHTON, MIFFLIN AND COMPANY
The Riverside Press, Cambridge
1895

Copyright, 1889,
By MARY E. BAMFORD.

All rights reserved.

The Riverside Press, **Cambridge,** Mass., *U. S. A.*
Electrotyped and Printed by H. O. Houghton & Company.

CONTENTS.

CHAPTER I.
Dredging Notes 1

CHAPTER II.
Water-Scorpions 17

CHAPTER III.
My Water-Lovers 30

CHAPTER IV.
Water-Boatmen 52

CHAPTER V.
Water-Tigers 59

CHAPTER VI.
Whirligigs 80

CHAPTER VII.
Water-Lizards and their Ilk 101

CHAPTER VIII.
Minor Mud and Water Folk 116

CHAPTER IX.
Caddis-Worms 135

CHAPTER X.

My Corydalus 144

CHAPTER XI.

Companions of my Solitude 154

CHAPTER XII.

Frogs, Boys, and other small Deer . . . 171

CHAPTER XIII.

A Lingering Good-Bye 203

NOTE.

The insects mentioned in these pages are those that I have found by hunting in several brooks in Alameda County, California. The fact of the distance of my point of observation from the places where most readers reside makes no especial difference, however, as members of the same families of insects will be found in or beside almost any brook, East or West. The various types of boy mentioned here probably exist in both sections of the country, also; and I will promise those who go dredging in the Atlantic-coast streams that there shall appear to them the apparitions of both the "fat woman" and the "wanderer from Erin's isle."

UP AND DOWN THE BROOKS.

CHAPTER I.

DREDGING NOTES.

> "Tush, tush! fear boys with bugs."
> *Taming of the Shrew.*

"You must n't *ever* let one of those big white dragon-flies come near you," said a little girl to me, impressively; "for, don't you know, they've got a needle and thread inside every one of them, and, if they catch you, they'll sew your ears up," and she looked at me with solemn childish eyes, evidently believing in the anticipated calamity.

"Yes," her little companion chimed in, "they'll sew your ears, an' eyes, an' nose, an' mouth up," and, having faithfully warned me, the little ones trotted up the bank and disappeared, leaving me to smile that the ancient prejudice against dragonflies should find such firm advocates beside a California stream.

Meantime, from the depths of the brook beside me, my muddy cloth dredger brings up various larval and perfect forms of insect-life. Great, sprawling, green larvæ of dragon-flies cling with their six legs to the dredger, and, to their indignation, are tumbled headlong into the pail that is to carry the findings home. Smaller larvæ of the *Agrion* dragon-flies come with them. Occasionally one of the *Hydrometridæ*, so-called water-spiders, or skaters, that spend life in an almost endless skate on top of the water, comes up in my dredger, gazes at me in surprise, and then skips back into the pool, to begin again the skating-match with his brethren, and to watch for any unlucky yellow morsel of a lady-bug that may chance to fall from the overhanging grasses into the brook. Did Don Luis Peralta, half a century ago, when he gave this land to Antonio Maria, know what a multitude of living creatures he gave with it?

Now and then one of the black *Dytiscidæ*, or Water-Tigers, an inch in length, tumbles clumsily from the dredger; and his smaller brethren abound. These *Dytiscidæ* are murderers at heart, as no one can doubt who has ever seen an earth-worm in their power. No sooner does the earth-worm fall into the water of the bottle in which these beetles

Water-skater.
Hydrotrechus remigis.

are confined, than one of the hungry *Dytiscidæ* will pounce on the unlucky creature. Another beetle, looking up from the bottom of the jar, will behold the prize to which his brother has fallen heir, and straightway, filled with covetousness, will rush upward through the water to pull the desired morsel away if possible. One beetle will tug in one direction, the other in another; they rush through the water, shaking their victim in perfect fury, till a person watching the battle might almost hear the first beetle squeak, " I will have it," and the other reply, " You shan't." And so the fight goes on, till one of the beetles conquers, and departs to enjoy the spoils of war.

Dytiscus.

Numbers of silvery beetles, — the water-boatmen, or *Notonectidæ*, — are brought to the surface, wrathfully skipping around in the dredger, and sometimes nimbly hopping back into the brook just as they are about to be transferred to the pail. Well do I remember my amazement, one day during my first acquaintance with these beetles, when, having transferred my silvery treasures to a pan of water, I had sat down to watch them as they swam on their backs, and, suddenly, one, the prettiest of the number, having turned over, flew straight out into the air, passed my ear with a booming sound like that of an angry hornet; and sailed away above the apple-trees, never to return. I have ever since retained a respect for the flying powers of the *Notonectidæ.*

Water-boatmen. *Notonectidæ.*

But my dredging is disturbed.

"What you catching? Fish?" demands a voice, and I look up to see the yellow head of an inquisitive fourteen-year-old youth peering over the bank.

Evidently he may have been watching my performances for some time, unperceived.

"Water-beetles," and I hold up my pail to show the contents.

"What are they good for?" proceeds the utilitarian.

I hesitate a moment. Shall I tell him of the decaying leaves that these numerous *Hydrophilidæ* devour, assisted by these pond-snails; of the yearly plague of frogs from which we are delivered by the disappearance of the juices of the polliwogs through the proboscides of these water-boatmen; of the multitudes of mosquitoes that never have a chance to bite us, because as wigglers they have met their fate under the masks of these dragon-fly larvæ? I excuse myself from this lecture on zoölogy, and make answer, "I take them home and keep them, and study their habits."

The boy eyes me suspiciously. Evidently that answer is not satisfactory to his mind. He thinks I am trying to cover up some great secret, some profound mystery that I object to his understanding.

He has constructed a theory of his own. Perhaps he lay awake nights thinking of the problem. He may have done so. I have been here often enough to excite the curiosity of any one who has observed me.

He has an idea. He will throw it at me and see if I, the guilty criminal, do not start when my secret plans are so openly disclosed.

He puts the momentous question.

"Do you make them into *medicine?*" he cries.

He eyes me as I repudiate the charge. I think he is persuaded in his own mind that he has hit on the exact solution of the mystery. I am a concoctor of horrible drugs. I fear that I am henceforth to be classed by him with those Chinese medicos who are rumored to concoct

Celestial medicine from horned toads and like crawling things.

Another large dragon-fly larva comes to the surface accompanied by two or three polliwogs. I remember being present at the massacre of a small company of polliwogs by some dragon-fly larvæ. The cool deliberation with which one of those larvæ, in particular, would crawl cautiously up a stick inside of the bottle, and, holding on near the end, would wait there until one of those black, immature toads came within reach, was terrible. Nearer came the polliwog, wriggling happily through the water, all unconscious of danger; and the larva, throwing out its mask and drawing it back again over the mouth so suddenly that the mask was just perceptible, would take a bite out of the polliwog. The poor victim would go rushing on, and the larva, having disposed of one mouthful, would pat its head with one foot, as if to pack the first morsel safely in, and then would reach out and take another bite from the next polliwog that came within reach, without reference at all to the fact that it was not the same polliwog to which the first mouthful belonged. Meantime, the other dragon-fly larvæ on the bottom of the jar were taking their meals in much the same fashion.

Large dragon-fly larva.

Such a sight gives one an idea of the multitudes of little tragedies that are enacted below the surfaces of ponds by these ferocious-looking larvæ and their victims.

But vengeance frequently overtakes the murderers. When the time draws nigh for moulting the skin and appearing with wings, then is the critical period of dragon-fly life. I recollect one wretch of a larva, that had spent his water-life as a blood-thirsty tyrant over the smaller creatures; but when the time arrived for moulting, he did not bravely crawl up a stick out of the water, and, seizing the end of the stick with his six legs, proceed to make an opening in the upper part of his back, and come out of that improvised door, after the common manner of dragon-flies. He seemed to be in a very excited frame of mind, climbed the stick, tumbled off on the floor, and crawled vigorously around in all directions, evidently in great trouble. At last, after half a day or more of such excited actions, he did manage to break a way through his casement and come forth; but, alas, he was never able to straighten out more than one of his four wings. The others remained immature, or wrapped around his body, and all his efforts were unavailing. He died in the struggle, and the ants were his undertakers.

None of the smaller dragon-fly larvæ of the *Agrionidæ*, that I have ever seen make their entrance into the air-world, have had such trouble as the larger ones seem sometimes subject to. The moult is often over, and the dragon-fly is ready for flight in an hour or two from the time of the beginning of the performance. It is easy to know when one of the larvæ is about to moult, since, for a day or less before this event, the larvæ are in the habit of crawling up the stick that is always left in the jar to serve as a sort of ladder, and of staying near the surface of the water, occasionally putting their heads out into the air. It is an interesting sight to watch the dragon-fly, after moulting, when the wings are gradually being drawn out to their full size, the fine veinings slowly becoming more and more distinct, spots of green or blue, markings of brown and yellow, or shades of pink and straw-color are making their appearance, while the dragon-fly occasionally lifts one foot and passes it over its head, moving the joint of the neck and bobbing the head up and down, as if to be sure that it is securely fastened on, and has not become loose in the pulling off of that skin overcoat.

Once in a while a dragon-fly makes a mistake, and leaves one leg behind him in his haste to get out of his old dress; but there is no going back and looking in pockets for anything that may be missed. Such a dragon-fly is henceforth five-footed, and seems to suffer but little inconvenience from the lack of the sixth member, except that in crawling there appears to be an inclination to tip slightly toward the side that has not enough support. The small dragon-fly larvæ seem to be a fraction less ferocious than the large ones, and to them we owe a debt of gratitude, since they make a point of consuming as many mosquito-larvæ as can be obtained.

The mask of the dragon-fly is somewhat like that ancient implement known as a " catch-poll," wherewith officers of the law were wont to apprehend criminals who were taking to their feet for liberty. The poll-part of the instrument being about six feet long was fitted with an ingenious steel apparatus at the end consisting of two V-shaped arms and a collar. The arms being flexible the head of the offender was allowed to slip past them, but when the collar was once around his neck he was securely held, and then the officer calmly dragged his victim to prison, or pushed him there, as suited the offended administrator of the law. Some collars of the " catch-polls" had rows of sharp spikes set round them, so that

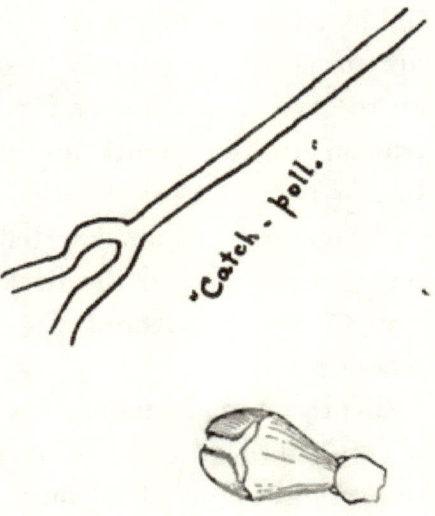

Mask of *Libellula* dragon-fly larva.

if the offenders struggled they might be hurt. But the dragon-fly larva's "catch-poll" disposes of the offender at once.

Woe is me! A piping voice cleaves the air.

"You catching fish?"

Verily, the nature of the small boy has not changed much since Wood wrote: "At the best of times the microscopic angler is sure to be beset with inquisitive boys of all sizes, who cannot believe that any one can use a net in a pond except for the purpose of catching fish, and is therefore liable to have his sanity called in question, and his proceedings greatly disturbed." Wood goes on to give as a remedy for this evil the administration to the small boy of "soft-sawder and a few pence." Perhaps the California boy is not so avaricious as his English cousin. At least, I have usually found the first half of this prescription sufficient, without the administration of the second.

Truth compels me to state, however, that some of my small allies have assisted me with the not improbable hope that some little fish might come up with the beetles in the dredger, and that these fish might become the property of the boys, who are prone to have wild hopes of raising fish in tin cans, — albeit such schemes usually end in the fish's living a few days on bread-crumbs, and then expiring, to the owner's grief.

But the small boy, when once enlisted in the

work, becomes a most enthusiastic ally. In fact, he soon ceases to occupy that position, and becomes commander of the expedition himself. So very enthusiastic does he become at times, that he splashes around in such a manner as to impress even the most stupid kinds of bugs with the idea that danger is near, and consequently they seek their hiding-places with such rapidity that search after them sometimes is useless, in spite of the small boy's well-meant zeal. This individual is useful, however, in reaching for specimens, which can be had only by standing in muddy places, or on precipitous declivities where a woman might find difficult footing. I owe at least one good specimen of a water-inhabitant to a small boy's zeal — a fact that I have often thought of with compunction, inasmuch as, just before receiving the specimen, I had publicly reviled this same small boy as being one who intended to keep my dredger all day and allow me no use of it whatever. But the three brown-headed intruders who now plunge down the bank have urgent business on hand.

"What are you after?" I ask, by way of return catechism.

"Red-legs," responds one freckled urchin, making a dive into the brook, and on being questioned further, it appears that this is the name of a species of frog that the boys hope to sell for a fabulous sum to a mysterious Frenchman. But no such frog appears, and they run on farther up the creek to continue their search.

Here and there, on the leaves or chips that float in the water, or on the grasses that hang into the stream, one finds clear, yellowish jelly drops, as big as dew-drops, or perhaps larger. To those who have kept these drops and seen their final outcome, they are known as the eggs of the common pond-snail. Dear to my memory is the first little pond-snail that ever hatched in my own bottle of snail-eggs. How eagerly from day to day I had gazed into the depths of the water in my bottle, hoping amid the grains of sand and specks of leaves to see some movement indicating life, and how rapturously, on the twenty-sixth day of my search, did I see through my microscope the motion of a little thing about half as big as a pin-head. The speck grew, and behold, it had a wonderful little shell, and, at last, one day I attained the height of my ambition in snail-culture, for I scraped the clinging sand from the minute object, and there was revealed to my admiring gaze the baby snail, the little black whorls of its tiny shell as perfect as those of the biggest of the family.

There are three kinds of snails that I have found in this brook; the "left-hand," the flat, and the "right-hand" snails. They all have the same meek, quiet nature. In regard to their appellations, if you hold the shell of one of these snails with the apex up and the opening facing you, that opening may seem to be either on the left or on the right; that is, it may be either a sinistral aperture or a dextral one. If the former, the snail belongs to the *Physa* division. You can see such snails floating wrong side up in these pools, taking a swim after their own peculiar fashion. I think that while on these swimming excursions the snails are sometimes brought into contact with the polliwogs. I saw a pond-snail swimming once when a polliwog was very persistent in his attentions to it as it floated on top of the water. A boy who was with me was much interested in the performance, thinking, I suppose, that the snail was about to be eaten.

The snails themselves (I am now speaking of *Physa*) are black, or very nearly that color. One can observe the flesh when the snail is swimming or walking. When it is dead its shell is of a lighter shade, a grayish or yellowish color.

The method of swimming has been variously described as " clinging with their foot to the surface of the water," and as a " creeping on the *under surface* of the air," that is, on the layer of air next to the pond. It is a queer performance to watch, for the shell and the creature in it look weighty enough to sink to the bottom of the pool, yet the swimming or floating goes on.

There are much smaller snails in this brook. One finds them only occasionally, at least at certain seasons of the year, I think. They are the flat ones that have all the whorls of their shells in the same plane. These snails belong to the old genus Planorbis.

It is not particularly wise to leave a snail-bottle open. Close it with mosquito-bar, for otherwise you will be likely to come back after leaving your bottle and find out that your biggest and therefore supposedly your most intelligent pond-snail has walked out of the jar, deposited himself on the steps in the hot sun, been unable to extricate himself from his difficulty, and find his road back to the water in the bottle, and so has perished, miserably broiled to death. So much intelligence has the pond-snail.

It is said that, in Rome, the present of a snail on a certain festival was a symbol of renewed friendship. I can well believe it, for although I suppose it was a land-snail, yet snails whether on the earth or in the water are most peaceable creatures, and usually set a much better example to the quarrelsome beetles than those creatures are willing to follow.

Pick up any stick that you may find that has lain long in the water, and you can gather pond-snails from it. They hang on the water-weeds also, where those dip into the pools, and if the snails are gathered and kept a few days in the spring, their eggs will be found on the sides of the jar.

Yet the fact of the matter is that there is not much character to a pond-snail. To slip out of a mass of jelly with one's house on one's back, to float on the surface of a pond, to dine on leaves or confervæ, to rest when weary and to journey when so disposed, to retreat into one's house when in danger, to pass along through life in a somewhat humdrum fashion with small spirit or vim in one, to cleanse the pool the little one may, and finally to drop down through the water and whiten with one's lifeless shell the slime of the pond, to have that slime close at last over one's shell and leave one buried in oblivion while all the pond-life goes on above one still,—this is a snail's life. Devoid of fighting instincts, not gifted with ambition to soar like the beetles, or to be ever in sight like the skaters, treating all the pond-neighbors with quiet reserve, going one's own way and doing much good in the world, such is the pond-snail. If he is not brilliant, he is good, and what more could be asked of him?

There is a stir in the grass at the top of the bank.

"What do you s'pose she's getting?" says one low voice.

A bug-hunter learns to listen and generally hears much that is said about dredging.

"Fish, I guess," answers another boy, confidently; and they pass on, leaving me to climb the bank and wend my way homeward, battling all the time with two or three obstreperous water-boatmen, that with buzzings of defiance are endeavoring to climb the sides of the pail and take flight back to the brook from which I have just drawn them.

CHAPTER II.

WATER-SCORPIONS.

"I do desire we may be better strangers."
As You Like It.

A SMALL boy stood beside the brook, gazing intently at my exertions. His cheeks were very rosy and a wide patch of mud adorned either side of his mouth. The brook evidently had attractions for him, but he was dissatisfied with the present results of dredging.

Water-scorpion.

"Say, did you ever catch them kind of bugs that look like shrimps?" asked he, after eying my bottle. "Not just like them, either. There's one, now" — and the dredger came up with an ill-looking scorpion-bug clinging to it.

"One of them fellers caught hold of me once," he went on, confidentially, — "right hold of my thumb-nail, and he just held on and pulled till my thumb-nail came out. It's growed again, though," and, in proof of his thrilling tale, he held up to view a thumb that certainly showed no trace of the dire combat described.

Scorpion-bug bearing epaulette, — rather enlarged.

"What did you do with the bug?" I asked.

"Killed him," responded the youth.

When bumble-bees buzz round the blackberry blossoms in April, occasionally one finds in this brook a female scorpion-bug bearing her eggs on her back, looking as though a second story had been built on top of her, the egg-mass being sometimes as thick as the average scorpion-bug's body. The eggs are placed regularly, standing up from the wings. On one such bug's back I counted one hundred and twenty-six eggs. Perhaps I was slightly mistaken, as the work of counting was somewhat difficult, but that must have been about the number.

A bug bearing such a burden as this is easily scared. Solitary confinement literally frightens the creature to death. Twice I have attempted to keep such a bug alone, but within twenty-four hours or so the poor prisoner would be found floating, dead. **Mrs. Scorpion-Bug** evidently objects to being monarch of all she surveys.

Some scorpion-bugs carry merely a few eggs, and do not cover their backs entirely. One of my scorpion-bugs used to go wandering around the bottle carrying half-a-dozen eggs or so as a kind of epaulette on her right shoulder, the rest of her back being bare. A full covering of eggs makes a scorpion look as though she had the uneven, brown outside of an acorn-cup on her back. One finds these cast-off egg-foundations in the water. The eggs do not seem to be very tightly fastened to the bug's back, and hitting against other beetles or pieces of wood loosens the egg-mass till it comes off whole and floats in the water.

If one of these egg-burdened bugs dies, the egg-mass may easily be separated from her back, but I do not think that the eggs will hatch. At least, my experiments in trying to keep such eggs till hatching have proved unsuccessful. Occasionally I have found in the jar on a stick eggs of the same shape evidently deposited by other scorpion-bugs. The baby scorpion-bugs are miniatures of their parents, looking a little like a lot of infant squash-bugs that had suddenly taken a fancy to swimming. Once in a while one will see a youthful bug catch a tadpole half as big as himself. Mosquito-wrigglers form part of the bill of fare also.

But the lives of the infant murderers are generally short in my bottles. The young scorpion-bugs stay about the surface of the water, heads downward, and the bigger bugs evidently consider them as so many joints of meat hung up at the butchers to be hastily disposed of by customers. The baby scorpions must do a deal of dodging about in the brook.

I believe that it is Herodotus who, among other truthful accounts of the Neuroi who once a year become were-wolves, and the Argippaioi who were bald and snub-nosed from their birth, speaks also of the Issedones who, according to his account, used to devour their dead parents with great pomp and ceremony. This order of things is quite reversed among the Water-scorpions; that is, the parents devour the children, but there is no pomp and ceremony about the performance. And from all I have seen I have no doubt that the infants would willingly follow the custom of the Issedones if size would permit.

The large scorpion-bugs, when kept in a bottle, have a habit of choosing some chip or bit of wood and using it as a raft, half-a-dozen or so of the bugs climbing upon it and going sailing up and down from the surface to the bottom of the water, back and forth, on their improvised bark. One does not see the other water-inhabitants taking such rides, as a general thing. The only exceptions that I remember are the "Whirligig" beetles, and their rafts are as much smaller as they themselves are smaller than the scorpion-bugs. Other water-creatures crawl over the pieces of wood, to be sure, but they do not seem to care about sailing. It seems to be an amusement original with the scorpion-bugs.

If one comes upon the jar suddenly, one will see the scorpions making a hurried descent to the bottom of the jar, their wicked-looking front legs thrust out before them. Arriving at the bottom and becoming assured almost immediately that there is no real danger, they seize something, perhaps a cast-off egg-foundation, and rise with it to the surface of the water once more. Scorpion-bugs have a habit, when sitting or hanging on a piece of wood in the water, of passing the front feet over the head, as one sees a fly do.

These water-scorpions have often a peculiar sign that precedes their death. A few days before that event the body gapes wide open, as though the creature had been sliced by a knife which had been passed horizontally under the wings, between them and the body, leaving the two pressed widely apart. This gaping of the body is generally, though perhaps not always, a fatal sign, and it is certainly a very noticeable one. When the cold days of winter come, one will sometimes find the bottom of the jar covered with dead scorpion-bugs, their bodies gaping wide open.

But words cannot describe the horrifying appearance of lively scorpion-bugs. One might well think a company of them to be a sight of the "nightmare and her ninefold." However, I think that most of their threatening actions are merely expressions of fright, they being as anxious to escape as most people are to let them go. But this variety, the length of which is only about thirteen sixteenths of an inch and the width six or seven sixteenths in the broadest part of the back, becomes insignificant when one thinks of its

Belostoma grandis.

near Brazilian relative, *Belostoma grandis*, a full-grown specimen of which is about five inches long, the wings when spread measuring from tip to tip more than seven inches. Methinks it might require some courage to explore a Brazilian brook, since one is apt to give a nervous jump now and then even after one has kept these small scorpion-bugs in captivity for months and has become somewhat accustomed to their alarming appearance and actions.

The water-scorpions of this brook do not possess the thin breathing-tube with which the bodies of some (*Nepa*) of this family terminate. I found in this water, however, one tube-bearing member of the *Nepidæ*, a Ranatra. As all know who have found this creature, Ranatra is long and lank, looking like a quite thick black darning-needle that had taken to itself six bent pine-needles for legs and two more for breathing-tube. My Ranatra was about two-and-a-half inches long, and his thin self bore no resemblance at all to his cousins, the flat, broad, brown scorpion-bugs.

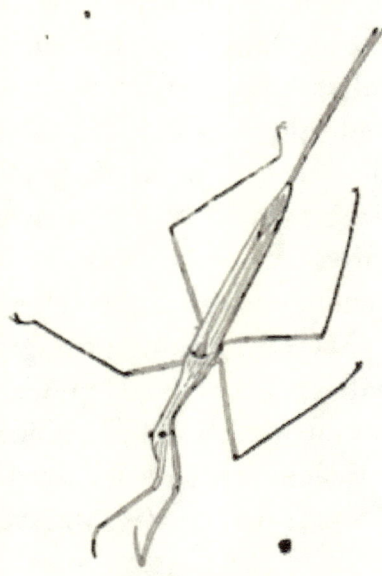

My Ranatra.

Poor Ranatra! "The best in this kind are but shadows." It was difficult for people, on his first acquaintance, to believe that he was really "done." He looked more like the skeleton framework of an insect than like a finished bug. I remember once having explained the metamorphoses of the larvæ of a water-beetle and of those of dragon-flies to a looker at my captives, and, at the end of my zealous discourse, the listener pointed through the glass at poor Ranatra and said, "And what does that one turn into?"

In Wood's "Trespassers," the naturalist mentions a Ranatra that was caught by a girl and named "Daddy," from its resemblance to a Daddy-long-legs fly without wings. The naturalist goes on to say that Daddy and the girl had fights together, in which the girl irritated the insect with a pencil, and Daddy struck back again. "The courage of Daddy and the force of the blows which he delivered on the pencil were well worthy of notice."

My mind being wrought up by this account of battle, I resolved to wage war boldly with my Ranatra, also. The choice of weapons being mine, I took a white stick and proceeded to make all manner of passes and raps at Ranatra. But I grieve to state that Ranatra was a coward. This was certainly true. He fled dismayed from the combat, and, standing upside down, endeavored to hide his head in a hole. Perseveringly I chased him from his hiding-places, but not a blow would he give back. Peaceable creature! The only time that I ever knew him to manifest what could be construed as, perchance, gestures of meek-mannered anger was when I prodded him maliciously until he was forced to let go his hold on a beetle, the life-juice of which he was in the raptures of sucking. This bold interference with his dinner-rights caused a few clawings with his forelegs, but that was all.

Ranatra's long breathing-tube was quite flexible, being attached to the body in such a way that he could allow some stick or leaf to bend it at quite an angle, until one dreaded the next minute seeing the tube snap off, and wondered what in the world Ranatra would do then. Yet no accident ever happened.

His second and third pairs of thin legs were fringed with very fine hairs, scarcely noticeable without the microscope, unless seen in the right light. He could hold up his fore feet in the same reverential attitude that the land-insect, the Mantis, employs when waiting for prey, when, as old De Mouffet said, " It alwaies holds up its fore feet like hands, praying as it were after the manner of the Diviners, who in that gesture did pour out their supplications to their gods." The attitude had about as much religious significance in my Ranatra as in the Mantis. Small dragon-fly larvæ (*Agrion*) were tid-bits for Ranatra. His rounded head with its sharp bill was very suggestive of that of a minute bird. His eyes were very prominent, standing out in round, black balls upon his head.

If ever a bug was a hypocrite it was Ranatra. I have seen him when a water-boatman would come and perch on his back. Ranatra would be perfectly quiet. He knew he had been taken for a stick, which he so much resembled, and a stick he would be. And when the water-boatman suddenly took to its oars once more, not the slightest motion showed that Ranatra was alive. He would hang poised over an unconscious bug, like Damocles' sword, but, though long delayed, at some sudden moment, the sword became alive, and piercing the unfortunate bug finished its life.

I think Ranatra had no music in his soul, and he probably never missed the bird-twitterings of his native brook. As a personal favor and a reminder of the days when he lived in the creek, I sometimes took a flute and played "'Way down upon the Swanee River" close to his jar. But the calmness with which he received the serenade was only equal to that with which he usually surveyed the world when no music was going on. Neither the "growly" nor the "squeaky" parts of the piano affected his nerves, even when his bottle was placed touching the instrument next the keys. It was fitting, however, that music should have no charms for such a deceiver as Ranatra. Does not Shakespeare say, —

> "The" (bug) "that hath no music in himself,
> Nor is not moved with concord of sweet sounds,
> Is fit for treasons, stratagems, and spoils?"

For all of which Ranatra was most eminently fitted. Many a time had I thought Ranatra dead, but the rascal came to life a few minutes afterward. So, one day, when I had had him in my possession about nine months, I hardly believed him dead upon finding him lying prone in the water. He had shammed so much that I hustled him around shamefully before I believed the fact of his decease. But Ranatra was truly dead this time. I put him in a separate jar of water, having a faint hope that he might revive during the night, but in the morning he was still limp, and a couple of pond-snails, one on each side of him, were performing the last kind office for the dead in cleansing Ranatra from the green scum that had attached itself to him.

Peace to his ashes. I did not know how I loved him until he died. Never did I part from a bug with such regret. No post-mortem examination that I could have made would have revealed the cause of his death. Perhaps it was old age, since he was fully grown to all appearances when I found him, and wise men tell us that the life of an insect is often not much more than a year in length. It may be he was Ranatra the Aged. The jar looked lonely without him, he had lived in it so long, and I felt half inclined to think that, in spite of his having dwelt with them so securely for so long a time, he had at last fallen a victim to some of those cowardly cousins of his, the scorpion-bugs. They rushed about as usual, evidently caring nothing for the death of the bug that was worth twenty of them.

Even in death Ranatra kept those fore feet held up in their customary reverential attitude. It was enough to excite superstition, if a superstitious person had seen him. The Rabbis have said that locusts were made out of the superfluous earth that was left over after the making of Adam. Mayhap if the Rabbis had known Ranatra, they might have told some like marvellous tale of him. The Rabbis were gifted with originality of statement, to say the least, for did they not say that, when Adam wept after the fall, though every beast and bird hastened to mingle their tears with his, yet the locust arrived first, on account of its kindred origin with Adam? And, furthermore, are not the black marks on the locust's wings Hebrew characters, wherein may be read this: "God is One; He overcometh the mighty; the locusts are a portion of His army which He sends against the wicked"?

To be sure, Ben Omar says that the prophet Mohammed read the Hebrew characters differently, for he made them out to signify this: "We are the troops of the Most High God; we each lay ninety-nine eggs. If we were to lay a hundred we should devastate the whole world." On account of the great alarm that these mysterious words raised in the mind of the Prophet it was granted that an invocation addressed to Mohammed, and written on a piece of paper enclosed in a reed planted in the midst of a wheat-field or of an orchard, has the power to turn away the destructive hordes of locusts from the portion of the land so protected.

And when I recollect that the Rabbis were capable of making additions to the Scriptures so startling as that in the Targums which says that when the plague of locusts in the land of Egypt was disappearing before the wind, those insects that the Egyptians had fried or pickled and laid by in store for food were blown away from the land in company with the live ones, I am not capable of imagining what those Rabbis might not have said of **Ranatra**, had they known that singular-looking bug. I flatter myself that they did not have the pleasure of his acquaintance. Surely if they had they would have mentioned it. They usually mentioned everything they did know, and perhaps a little more. Ranatra did not realize how thankful he ought to have been that the Rabbis never saw him. Preposterous, indeed, would have been the story they would have tacked to his name. Ranatra made a beautiful corpse, calm and peaceful-looking.

Let no evil be remembered against him. He is dead.

"Bid him farewell, commit him to the grave;
Do him that kindness and take leave of him."

CHAPTER III.

MY WATER-LOVERS.

It is March. Let us go dredging for Water-Lovers, alias *Hydrophilidæ*. As one comes to the brook on such a day, one may glance across the ravine, and, suddenly, in between one's self and the vision of green fields and buttercups will come some fluttering wings, and a dainty-looking dark-brown butterfly with buff-colored margins, and a row of pale-blue spots inside those margins but too small to be noticeable at this distance, sits down on the opposite bank to sun itself beside the stream. The butterfly that our English friends call the Camberwell Beauty — *Vanessa Antiopa* — has come to look after our dredging. This creature flaps over our back-yards the last of February, and instead of appearing "with ragged and faded wings," as Harris says that this Vanessa does at the East, it here often looks as fresh and new as if just made.

One day I heard a sad tale about this Vanessa from two young sinners of twelve years old or so that I found enjoying a solitary cigarette between

California Hydrophilidæ.

them in the seclusion of one part of this brook. The lads were mightily interested in my dredging, albeit they looked at me as though they expected that I would begin a sermon on the evils of tobacco. But what is the use of scolding people who are so evidently aware of their own sinfulness?

I refrained from criticism, and was rewarded with a bit of information, false or true. A Vanessa of this sort flew by in the course of our conversation, and the boys called it a Japanese butterfly. They furthermore solemnly averred that the Japanese ate them. One of the boys even said that he had seen the Japanese do it, pulling off the wings and devouring the body. Whether the testimony of such youngsters is worth anything or not, I leave the judicious reader to determine, but certainly, if I were in Vanessa's place, I should keep as far away from both boys and Japanese as possible.

When March is nearly over, there comes an occasional Orange-tip butterfly to keep Vanessa company. In March, too, one passes by the little white-and-black kids that stand on the sides of the brook and dance along its precipices or stay in the safer green field by their mothers.

Burton in his "Anatomy of Melancholy" tells us that when Jupiter and Juno held their wedding-feast all the other divinities were invited and many noblemen besides, and among them came a Persian prince named Chrysalus, a very simple person, but gorgeously arrayed in golden attire. And the assembly, seeing him come with such pomp, were deceived, and rose up to give place to so noble a being. But Jupiter, seeing what a fantastic, idle fellow he was, turned him and his followers into butterflies, greatly to the astonishment of the guests, no doubt. If this be true, how comes it that Vanessa Antiopa wears black? Did Vanessa go to the wedding-feast wearing mourning?

It is a sad fact that butterflies are intemperate at times. To see the beautiful things flit by one would never think them guilty of such indiscretion. But a friend of mine informed me that just beyond these foot-hills he had seen one kind of butterfly on the buckeye-trees and could almost have taken the creatures with his hand. He supposed them to have been intoxicated with the fragrant flowers. I can well believe it, for, if I were a butterfly myself, it seems to me no flower would delight me more than the dense white panicles of the buckeye. Mr. Scudder gives a similar instance of the Tiger Swallow-tail butterflies being overcome with lilac blossoms.

If you look across the brook, you may see a man who fully realizes Bacon's assertion, that mustard-seed "hath in it a property and spirit hastily to get up and spread," for that laborer is gathering the yellow-flowered weed, and has a bundle of it at his side. Down by the water's edge the blackberries are in bloom, and in the grass on the sides are tufts of sorrel, reminding one of ancient Gerarde and his assertion that the wood-sorrel was of old called " Allelujah," or " Cuckoo's Meat," because, says the old herbalist, " when it springeth forth, the Cuckoo singeth most; at which time also Allelujahs were wont to be sung in our churches."

Here comes the first Water-Lover in the dredger. Once I tried to keep a colony of these *Hydrophilidæ*. They lived in a pickle-jar, and there were twenty-two of them. At least, after repeated attempts to take the census of the inhabitants of that jar, twenty-two was the number decided upon, it being somewhat difficult to count beetles that are in constant motion, and that are identical in appearance as far as one can see. Each beetle was three eighths of an inch long, one fourth wide, and appeared somewhat hump-backed. The upper, convex surface was black, as was the under surface of the beetle, but the latter usually appeared in the light as if covered with liquid silver, an effect produced by the air taken under water by the insects. The middle and hind pairs of legs were hairy, and all three pairs were armed with double spurs, some of them sufficiently powerful to feel, when applied to one's finger, much like a pin's prick.

Inspired by the gift of some weeds from their

native brook, the colony of twenty-two set up housekeeping. Unlike other beetles in neighboring jars of water, these twenty-two unanimously rejected the custom, fashionable among many water-creatures, of breathing through the abdomen, and, instead, on making their ascents to the surface of the water, put up their heads after the manner of *Hydrophilidæ* usually.

Although commonly silent these beetles were capable, when displeased, or when frightened by the danger of capture, of giving squeaks that might be heard across an ordinary room. These squeaks sounded somewhat like the shrieks of a fly when imprisoned in a spider's web, or like the croaks of a feeble-voiced toad. They were partially quiet when living by themselves, but from a jar of mixed beetles and bugs the shrieks were frequently heard.

Not long after the twenty-two set up housekeeping, there began to appear, on sticks and leaves in the jar, white egg-cases, as large as beans, or three eighths of an inch long, and a little more than one eighth in diameter, woven of white silk in a cylindrical shape and having sometimes a thin floating strip of white hanging from the top of the egg-cases. This strip was sometimes about five eighths of an inch long, and occasionally when an egg was taken out of the water the strip would fall back over the closed opening. One unacquainted with these eggs might think that only the wasteful beetles added this strip, the more economical ones refusing to expend silk

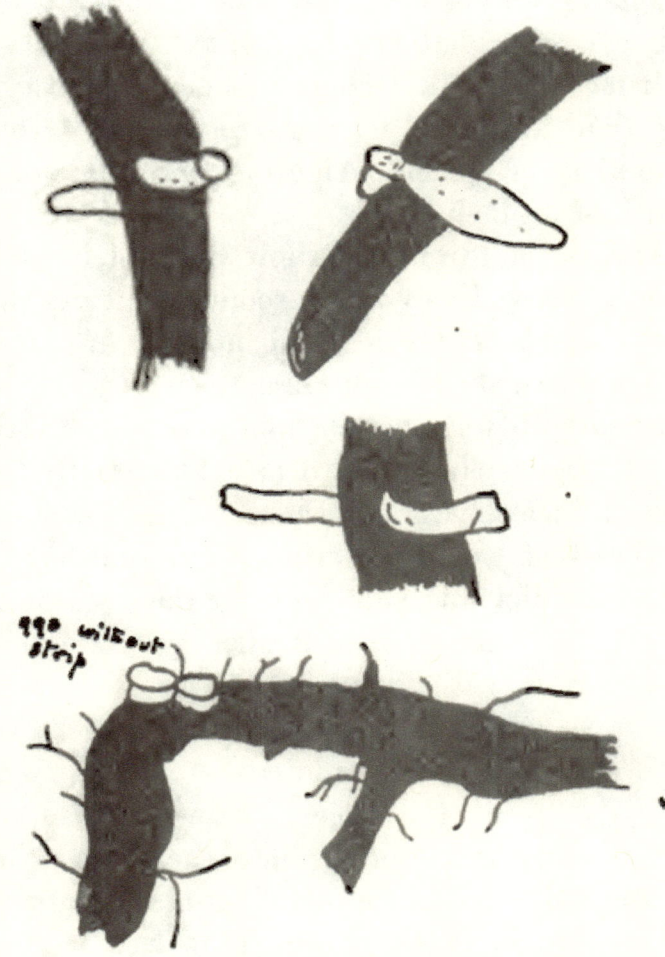

Eggs of *Hydrophilidæ* (Californian) as laid in my bottle. Eggs with and without floating strip.

in making an addition that was apparently of no use.

If some too-hasty mortal, wishing to perceive

what was inside of such an egg, opened it before the proper time, his curiosity was disappointed by finding only some little, yellow, oblong specks. The eggs often hatched in about thirty days from the time of laying. There issued from the first egg that appeared in the *Hydrophilidæ* bottle a number of queer, wriggling larvæ, about one eighth of an inch long.

And as children will show what their parents were, so these lively larvæ soon let out the secret that the black beetles had not by any means always pursued the noiseless tenor of their way, satisfied with water-weeds and an occasional nibble at a departed earth-worm. The family traits showed themselves, and the *Hydrophilidæ* stood convicted of no less a crime than murder. No vegetable diet was sufficient for the appetites of these little squirming larvæ, that, with their branching pincers projecting from either side of their heads, wandered around the surface of the water, occasionally meeting and bunting into each other, interlocking horns like a couple of inimical goats.

Larvæ of California *Hydrophilidæ*.

Those little black worms that have a habit, when confined in bottles, of inching up the sides of the glass until above the water, had great enemies in these larvæ. These worms are the larvæ of a variety of gnat, and their pupæ, instead of being lively, like the "tumblers" of ordinary mosquitoes, are quiet, black, round, and rather hard.

No sooner did one such worm descend from his height in order to make a voyage from one side of the bottle to the other than his path was beset with danger. The horns of the larvæ pointed toward him, and it became a question whether the worm would be a successful mariner or whether the dangers of the deep would be his undoing. If the latter, some pair of tiny nippers closed upon him, and the successful larva raised the victim aloft. But one larva should not be greedy, and his brethren all rushed to the feast, and soon the unfortunate worm was claimed by a number of larvæ, and, until the last vestige of that worm disappeared, the pulling and fighting went on. Often, however, the voyage ended successfully, and the worm would inch up the glass on the other side, leaving the larvæ wondering where he had gone to.

But murder of other creatures was not the only crime committed in that jar. Fratricide prevailed, until, at last, one larva remained, the sole ruler of the bottle, the survivor of his brethren, illustrating in himself the doctrine of the Survival of the Fittest.

The Fittest did not seem to be a very bright infant. Perhaps he was more deficient in intellect than *Hydrophilidæ* children usually are, but, at times in his career, I doubted whether, if he had been living in a brook and forced to seek his meals for himself, he could have succeeded in catching anything. I was his devoted slave and feeder. Everything he ate I put into his mouth.

It was not that the Fittest was not willing enough, but he was clumsy. He waddled, with the hinder part of his body held above, much after the manner of a misdirected parasol, while his six legs paddled through the water, and his horns turned stupidly toward any prey that he dimly saw. He had caught his brethren because they were as stupid as he, but many things were quicker. Mosquito larvæ utterly foiled him, darting by before he could begin to turn his horns in their direction.

The black-headed worm had a trick that always completely fooled the Fittest. No matter how vigorously that worm had been wriggling, if it came next to the Fittest's horns it remained perfectly still; and the Fittest, with that sublime stupidity that characterized him, would pass that worm by as though it were not alive. In fact, I began to think that the only thing the Fittest did perceive was motion; shape was beyond his mental powers. If a thing wriggled, he thought it was food; if not, it was nothing of interest to him. The black-headed worm was infinitely ahead of the Fittest in intellect.

The only way I made sure that my captive had any food was to take a straw and carefully steer one of the black-headed worms to the harbor within the Fittest's horns. Then, feeling my straw tickle, the worm would forget its caution and wriggle directly before the Fittest's little black eyes, and then he saw at last the meaning of my efforts and was soon waving his victim in triumph.

After one of these enforced meals, the Fittest descended on a day to the bottom of the water in the tooth-powder dish where he resided and indulged in a season of meditation. He disdained to eat, and only occasionally varied his meditations with strolls around the floor of his home. These strolls exhibited one of the Fittest's accomplishments; he could travel backward as well as forward. He was also as flexible as the Japanese boy one sees at acrobatic entertainments; he could make a perfect ring of himself. Did some invisible speck of dirt attach itself to his tail during his promenades, all the Fittest had to do was to bend his head straight back over his back, take hold of his tail, and, straightway removing the trouble with his pincers, go on with his pilgrimage in peace.

Harmless as the Fittest was, he must have had some terrors for a minute May-fly larva that dwelt in the same tooth-powder dish. The little May-fly developed a habit of giving a rapid skip and hopping over the Fittest whenever they met. But, alas for the Fittest, not even a May-fly-larva need fear him long. He drooped; black-headed worms lost all charms for him; he travelled neither backward nor forward; grim Death claimed him for his own, and a prospective beetle was lost to the world.

The *Hydrophilidæ* themselves had their own troubles. Opening the jar one day, I found two of the beetles carrying a yellow burden apiece. On examination the burdens proved to be small yellow leeches that had calmly appropriated these two beetles and forced them to carry them around on their backs. Securing one of the beetles I attacked its burden, but it was only after much pulling and many exertions both by the beetle and myself that the leech finally let go and collapsed into a disgusted heap in one corner of the spoon. During the struggle a number of baby leeches had detached themselves from their parent and were scattered over the spoon. On putting the leeches into water the children gradually began to come back and cling once more to the larger leech. I had a similar struggle with the leech attached to the other beetle, but the young leeches were smaller than those in the first instance. Think of the sorrows of a poor beetle in a pond, with no friendly hand to take such a life-menacing burden from his back.

After diligent search, another larva was fished from the pool in which the *Hydrophilidæ* had their former residence, and was brought home to fill the place of the Fittest. This the larva could more than do, as far as size was concerned, being at least four times as large as his predecessor. The surroundings of this larva were more favorable than those of the Fittest had been, inasmuch as on some water-weeds in the jar of the new larva were a number of the small crustaceans allied to beach-fleas and known as "fresh-water-shrimps," which were to serve as the larva's destined prey. Their often slow movements were just suited to the larva's mind. There was no such levity about them as about a black-headed worm. The larva could walk carefully along the weed, and, coming upon a water-shrimp unawares, grasp it with those pincers of his and devour his prey. He would elevate the shrimp, above his head out of the water, and in some mysterious manner manage to keep his tail out of water too, in order to breathe, while the main part of his body was under water; and, in this position, he devoured his meal, moulding the water-shrimp between his pincers as one might a bit of gum. One water-shrimp that I timed him on was eaten in seventeen minutes. His appetite for these creatures was good. In a little time after eating one, this Devourer wanted another. Perhaps a taste of this congenial food might have saved the life of the Fittest. Howbeit, water-shrimps are not all of life, and the Devourer must have longed for something more, for "one morn I missed him on the accustomed " weed, nor yet upon his stick nor in the jar was he. Had he fled? The mosquito-bar over the mouth of his jar showed no trace of his passage. Had he departed this life? Diligent search among the débris in the bottom of his jar failed to bring to light his corpse. The fate of the Devourer will ever remain a mystery. Sufficient was it to know that this second attempt at raising beetle was a failure. There was, however, a rumor of the assassination of the Devourer by water-shrimps.

A great many books of reference speak of the *Hydrophilidæ* as entirely herbivorous when mature beetles. Dead flesh, however, hath charms for some of these reputedly strict vegetarians. I have seen one of these *Hydrophilidæ* so interested in a dead earth-worm that when his supply of air became exhausted he would rush to the top of the bottle, poke up his head, rush down again and go straight to the worm and recommence the apparent chewing off of little pieces of flesh. However, the *Hydrophilidæ* get along very well when shut up in a bottle with nothing but weeds, but I believe that when at home in their native pools they sometimes act as scavengers, not only in the matter of decaying leaves but in that of dead flesh as well.

Professor Karl Semper, of the University of Würzburg, in speaking of the food of animals, mentions the European pond-snail, *Lymnæa stag-*

nalis, as often turning from its plant-eating to prey upon little water-tritons, attacking quite healthy specimens, overcoming and devouring them. Perhaps it may have been some such rebound from the common habit of life that influenced the beetle I observed to illustrate Professor Semper's assertion that "many polyphagous species are found in genera which otherwise contain none but monophagous Carnivora or Herbivora."

The Water-Lovers go armed. At least it has been conjectured that they do. Concealed weapons are not forbidden in Water-Land. Safely hidden underneath the body of one of these *Hydrophilidæ*, at the end of the sternum, is a very short, somewhat sharp, black "pin," ready, it might seem, to be stuck into the unfortunate creature that incurs this beetle's wrath. An experienced youth whom I once discovered, or, rather, who discovered me, during my dredging-hours, announced to me that the *Hydrophilidæ*, or "toe pinchers," as he called them, "bit."

"How do you know?" I asked.

"Got them on me when I went in swimming," responded the wise youngster. Perhaps he had felt the thrust of this same black "pin." But lads' eyes are not sometimes as bright as they ought to be, and I have quite a suspicion that the Water-Lover and the Water-Tiger beetles are confounded in the minds of many boys, since in this brook the most common varieties of these two families are much the same size, and to an unpractised eye are similar in appearance to each other. Of course, the bug-hunters, who know the two families, can distinguish them at once.

It is that way often in the insect world. The hunter after "bugs" sees at first the likenesses but not the differences. But when, as the reward of long and close looking, the latter become plain, the hunter wonders at not having seen them before. There is nothing like looking to sharpen the eyes, and the earlier one goes about it the more one will see.

Here come some more pond-snails from their watery homes. Let them swim while they may, for when these pools shall shortly dry up in the summer-time heat, then will the patient pond-snails, unable to take up their shells and fly away, do the next best thing, and, with becoming fortitude, burrow into the damp ground, hoping thereby to preserve their lives until water shall come again. Alas, rain will be months away.

Here, likewise, come up in the dredger some dark, perhaps two-inches-long leeches. There are plenty of them often in the mud at the sides of the stream. One can find numbers of leeches in autumn by taking a shovelful of mud from the pool's edge. Still, leeches are not delightful pets, and these may slip back in the pool.

Let us go home, for dredging is sometimes wearisome work on account of the strength it requires; but notice, as we go, how the tops of the green water-weeds are decorated with the white webs of these *Hydrophilidæ*. It is strange that, considering the number of eggs, the dredger does not more often catch the larvæ, but any one who has kept such creatures knows that they are very clever at concealing themselves. These egg-webs are clear above the water. If one did not know better one might pass them by as the webs of so many spiders. M. Figuier gives the time of the spinning of the egg-cases of the European *Hydrophilidæ* as April. Here the time is earlier, for one of my beetles celebrated St. Patrick's Day, March 17th, by spinning an egg-case. This is the earliest egg of these beetles that I have ever seen, although March 19th I found three in the brook, and one looked as though it might have been laid several days before. During the last ten days in March one can find many egg-webs on the reedy grass that dips into the water.

These *Hydrophilidæ* do not always place the eggs above the surface of the water, but, when these beetles are in jars, often put the webs where they are covered, and there is no air-tube rising from each egg to the air, as in the egg-cases of the large European beetles. If one drives away a beetle from her work before the case is closed one may look into the dainty white receptacle and see the bright-yellow egg-mass. A week afterward, if the beetle did not come back to finish her egg-case, one may look in and see that the egg-mass has separated into several oval eggs.

How full the world is of insect-life. Most of us have no idea of the number of our neighbors. It seems sometimes to the bug-hunter as though there would be but very few vacant rooms to rent in Nature's house. And yet the insects keep coming till even the flies take to the water. One can see a multitude of little flies sitting upon the surface of this pool sometimes. I do not know how many insects there are on an average to each plant in this country, but an English entomologist reckons that on an average there are six kinds to a plant in his country. Harris thinks that probably our average is less. Yet the hunter can see that there is hardly a crack or cranny in the bark of these trees but some insect has found it out.

Go by those willows in May and you will see leaves curiously connected, two or three fastened together near the tips. Tear such leaves apart, and lo, there is within a tiny gray weevil with pointed snout. Sometimes two or three of the weevils are in the same connected leaves. On those willows, too, in May, one may find numerous caterpillars, and the queer larvæ of what I suppose to be saw-flies, since, when touched, each larva has that strange habit, peculiar to saw-fly infants, of assuming curious attitudes, standing on the fore part of the body, the hinder part being held out in the air with the stiffness of a Sphinx caterpillar, excepting that the Sphinx holds out its head instead of its hinder portion. Sometimes one of these saw-fly larvæ curls its tail up, instead, much as a cat might curl up hers. This attitude is more striking than the other, if anything, for it looks almost unnatural to see the larva so twisted on the leaf. The larvæ hold on, during all their posings, by their fore feet. The color of the creatures is grayish on top, with ten double yellow marks on each side.

What a queer variety there is in insects! The more one sees of them the more one wonders at the marvellous diversity displayed in their appearance and organization. Come here next month, at the end of April, and, as you painfully pick out the gradually fattening larvæ of *Hydrophilidæ*, there will come a middle-sized, brilliant red dragon-fly whizzing almost into your very face. You may now find the larvæ of these insects, looking much like big spiders. A little boy who once saw one of them announced that it was a "grasshopper," but I think that the resemblance to a hairy spider is much more striking, the main difference, as one glances at the larva, being that a real spider has two more legs than this one. These larvæ, when disturbed, dig violently with their fore feet as if they wished to cover themselves with mud. However the larvæ are easily kept if fed with earth-worms, and I have had one transform into a red dragon-fly as early as March 11th. Unfortunately this red color is evanescent after death, much more so than the blue of the larger dragon-flies. The cast-off, spidery skins retain their shape, and around the opening that the dragon-fly leaves in the back on coming out are white threads, looking as though they were the bastings that held the dress together. "Wasserjungfern," "Virgins of the water," say the Germans when they speak of the dragon-flies. "Snake-stanger," or "snake-stang," is an old name suggested perhaps, as Dyer thinks, by the belief that the bite of the dragon-fly was as venomous as that of a snake.

Red Dragon-fly larva.

I do not know why the dragon-fly should have so bad a reputation. The larvæ are certainly hideous enough to frighten people, but it hardly seems as if the red, blue, or green dragon-flies need be much more frightful than so many brilliant humming-birds. I do not know what amount of horror might arise in the minds of those who now view the dragon-fly with superstitious fear did they know that the insect looks at them through compound eyes the facets of which are six-sided.

An army of teasels have marched down the hill in one place almost to the borders of this brook.

Now in March, when, amid the grass of spring, the dry, gray teasel stalks lie broken off or stand toppling to their fall, their brittle, hollow stems are still useful to the minor creatures. In many of the gray towers of the stalks abide spiders, clad in red and black, or the latter color only, sitting in grimness, like barons in the mediæval towers of yore, and intent on the same business that those gentlemen followed, the murdering of unwary travellers. Very frequently one will find a lady-bug with the spider, usually a defunct lady-bug, one of the kind that is red with no black dots on its back. Others there are with thirteen spots, and once I cut into a teasel tower containing a company of ants. If you look under the leaves of the fresh teasels which are springing up at the feet of the old ones, you may find perchance a snail, the hermit of this the Mediæval Age of Teasel Land.

Not always are the teasels friends of the insect tribe, however. These broad-based leaves form little hollows in which the showers deposit water in which unwary creatures are sometimes drowned.

Did you ever listen to the music that a crowd of dry teasel stems makes when moved by the wind? The sound is as musical as that one makes by blowing through a comb. Perhaps the spiders find such music lulling as the wind rocks them to sleep in the teasel towers.

Hordes of small caterpillars eat the teasel leaves

in June, and as one goes toward home carrying a bunch of leaves for such creatures, a white billy-goat near the path stretches out his head and mutely begs for a teasel leaf as a variation from his fare of dry grass. He munches the prickly morsel with satisfaction. Who had supposed that the teasels were the friends of so many creatures?

But when the teasels bloom, myriads of butterflies haunt the white-spotted heads. Here one will find little clouds of butterflies springing up before one while walking, and, if you look into the clouds, you may see what the boys call the Bull's-eye butterfly, *Junonia*, with its wings spotted with eyes like a peacock's. Here flutter blue or brown smaller ones, the representatives of the *Lycænidæ*, and beside this stream one day I found two downy, white moths, *Arctia*, with black spots on their wings, and abdomens marked on top with six black spots and a patch of orange.

In the summer time, when many of these pools are dry, one may walk their beds and find the butterflies resting and sunning themselves in spots where, a few months before, polliwogs bobbed up and down and pond-snails clung surrounded by dancing beetles. "Freyja's hens," the old Icelanders called the butterflies. "*Freyjuhæna*." Since Freyja was the Norse goddess of love, it is quite unlikely that a being of her sort would keep such good-for-nothing hens, the goddess being too much taken up with sentimental affairs to be much interested in poultry raising. All lovers who had been faithful unto death were afterward gathered in Freyja's halls, and, though the cynical might say that there probably would not be much of a crowd, yet even a small assembly of such ghosts would no doubt drive all thoughts of a practical nature from the poor goddess' mind. However, Freyja's hens are as useful as some real hens that I have had the pleasure of being acquainted with.

CHAPTER IV.

WATER-BOATMEN.

"I' faith, you are too angry."
Taming of the Shrew.

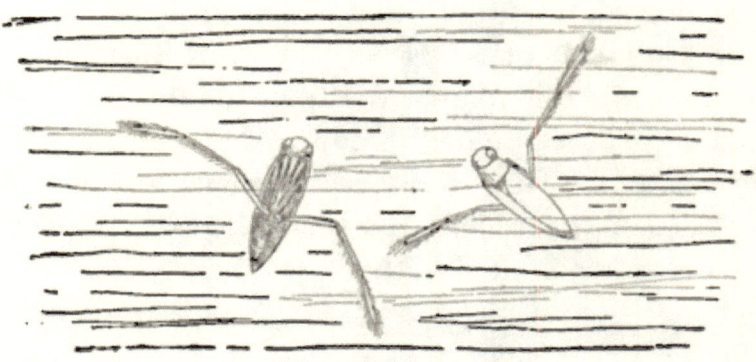

Notonecta glauca.

BUGS have tempers. I had as lief be a hornet as a water-boatman. The excitable disposition is just about the same. Buzz, jump, hop, defiance in every motion. One day, in my absence, a wayward cat thought to lie down to calm repose upon the steps where an open bottle of bugs stood. But alas for that cat's calculations! Either pussy had a day-mare during her slumbers, or else she sat down wrong, for thump! went the bottle, and the steps were deluged with a shower of " bugs " of all sizes and descriptions.

"Oh, how they hopped!" said the one who ran to the rescue, in relating afterwards the mis-adventure. "They acted as if they hated me."

And if, as I think may have been, the Water-boatmen were among the suddenly-landed "bugs," I have no doubt that they expressed their defiance in a series of hops that plainly told that they had "let their angry passions rise." Woe to the person that receives a prick from the proboscis of some of these creatures, for that person shall suffer as though stung by a wasp.

On a sunshiny day there is great excitement in the Water-boatman jar. Continued little clickings may be heard, made by the hitting of the bugs' heads against the sides of the glass jar. The surface of the water is surrounded with boatmen that have turned right-side up and are preparing their wings for flight. Every few minutes one of them soars upward, only to have his aspirations shocked by hitting against the mosquito-bar that covers the top of the bottle, and by being knocked back into the water, there either to try it over again or to disappear beneath the surface, where it is to be hoped the cool waves quench Notonecta's anger. At such times as this even calm Ranatra's disposition was corrupted by the evil world about him, and he elevated his head, and clawed at the glass above, as if he too would crawl out. Shall I set them all at liberty? How then shall I learn their habits? Do they then languish day by day, and pine in sorrow during their imprisonment? Nay, for does not Don Quixote say, "Sir, melancholy was not made for beasts but for men"? These very boatmen that now are so zealous for liberty will, by and by when the sun has gone behind that tree, sink down and solace these their woes by eating water-shrimps and earth-worms, and the jar-world will be calm once more. How many griefs doth eating assuage, not only in the bug, but in the human.

All people do not admire the boatmen. Well do I remember the frankness with which one urchin (who had his head bound up with a rag on account of an attack of poison-oak, and who found me gathering *Notonectidæ*) exclaimed, "If I was collectin' curiosities, I'd get something worth lookin' at!"

His charming directness of speech was equalled on another occasion that lingers in my memory by that of an old woman, whom I take to have been a wanderer from Erin's Isle staking out her goat, and who, looking upon me as I met her in the path, demanded what were the contents of my pail. Ignoring the suddenness of our introduction to one another, I brought up from the depths some of my beetles.

"Now," quoth the old woman, after a look at the uncanny beings, "how much nicer it would be for you to get fish than those things. There are plenty of fish in the brook, and anybody can have them," and she looked at me with a benevolently generous air, which, alas! was quite thrown away.

The eggs of the water-boatman I have found in March and July on sticks in my jar. These eggs are oval, white, and about one sixteenth of an inch long. They are sometimes in short rows, but often are scattered irregularly over the sticks.

One day I noticed on Ranatra's back an oval, white thing, and wondered what had befallen him. I fished him out and looked, and lo! it was an egg laid on his back by some much-deceived water-boatman that had taken him for a stick. Another egg

Eggs of Water-boatman in my bottle.

adorned one of his legs. Another time I found Ranatra with three boatman eggs on one leg, two on the leg opposite, and an old one on his back. Most likely these were the eggs of a lady boatman that was afterwards numbered among Ranatra's victims. How touching a spectacle to see the mother confiding to her murderer the future welfare of her children.

The infant water-boatman of the age of one or two days is very pretty and cunning, looking just like his ma, only very, very small. It is a fact well known to those acquainted with the *Notonectidæ* that there exist among the Water-boatmen different opinions as to the proper method of swimming. In the characteristic division, *Notonecta*, all the bugs consider it eminently proper to turn wrong side up, and swim in that humiliating posture; while the other division, known as *Corixa*, consider such actions unseemly and swim right side up with great propriety. Even from the egg these different modes of swimming are observed. It is in the blood, like any other family feud. Ranatra liked Corixa better for eating, a fact that did him credit.

No sudden start ever makes *Notonecta* let go its hold on its prey. Any unexpected jostle only sends the bug darting off, upside down, holding its prey firmly in its arms, as dreadful an embrace as a bear's hug. The two oars bear up the bug while it enjoys its meal. Occasionally *Notonecta* shifts its prey, and thrusts the pointed javelin of a beak into another place. The second and third joints of *Notonecta*'s oars are the most feathered ones, the hairs growing longer and thicker toward the end of the third joint, just as the blade of an oar becomes wider. Some Boatmen have a queer way of standing on the middle pair of legs, like a bug on stilts, the oars being spread in the water, and the short, first pair of legs clapping together.

Wood tells us that the eggs of a Mexican kind

of Water-boatman, of the variety Corixa, are eaten by the Mexicans under the name of "haoutle." The eggs are collected by sinking bundles of reeds in the water, and the Corixa lays its eggs on such bundles which are afterwards scraped and the eggs are made into cakes. It is well for the Mexicans that the Boatmen do not know of this performance. The Bohemians are said on the evening of all festival days to give some garlic to the house-dog, the cock, and the gander, it being thought that such food will make the three very brave. But I think the Boatmen would need no food to rouse the heroic spirit within them. If they heard of this Mexican performance, the next thing would be a pitched battle between the Boatmen and the supplanters of the Aztecs, and I fear the latter would hardly come off victorious. "Upon experience all these bugs grow familiar and easy to us," says L'Estrange. Perhaps they do. But I should be afraid to count too much on my familiarity with the Boatmen.

De Geer supposed that the Water-boatman drops some poison into its bite, since insects die so soon after being pierced. And Wood likens the dull, aching pain that follows the smart prick of the proboscis of Corixa to the sting of a wasp after the first sharpness of that has passed. I cannot myself state how it feels, for I am thankful to say that, in all my acquaintance with these bugs, I have escaped their proboscides. I am quite sure, though, that they would have been most happy to have afforded me the sensation of being penetrated by their beaks, and their only regret in that case would have been that they could not at the same time eject a bit of the *acqua tofana* of Perugia into my veins.

CHAPTER V.

WATER-TIGERS.

"Let's talk of graves, of worms, and epitaphs."
Richard II.

In April one may find in this brook the nearly full-grown larvæ of those beetles known as the Water-tigers, or *Dytiscidæ*. These larvæ are ferocious creatures, as the children of water-tigers well might be. They are strong and slender, furnished at one end of the body with a flat head marked with six ocelli and armed with a pair of sharp jaws like scissors, and at the other end by two breathing gills

Larva of *Dytiscus marginalis*.

which they keep uppermost as they dart head downward through the water. Armed with his pair of shears, the gray-yellowish, two-inch-long larva goes forth to prune the animal world. Is that a polliwog? Let us snip off his tail. It is too long. Or, if that cannot be accomplished, let us at least hold on to the polliwog till we have sucked him dry of juice.

I do not think it is possible for two of these water-tiger larvæ to live together in the same bottle. A battle will sooner or later occur and one will be killed. The indiscriminate slaughter of victims indulged in by these larvæ soon imparts to an uncared-for jar an "ancient and fish-like smell," since the larvæ do not devour their victims whole, but suck out the juice and then drop the bodies on the bottom of the jar. A keeper of these larvæ will find himself called on often to perform the office of undertaker.

But the sickle-jaws do not always prove all-powerful. There are individuals that refuse to be pruned by such a pair of shears. A big dragon-fly larva is a match for the shears-bearer. I left one of the *Dytiscidæ* larvæ once in a bottle of large dragon-fly larvæ, and when I came back the shears-bearer was himself divided into two parts. The dragon-flies had conquered.

Like some of the dragon-fly larvæ when they are about to enter upon the struggle of coming out with wings, some of these beetle-larvæ think that "the melancholy days have come" when the time approaches for leaving the water and burrowing in the ground for the pupa-sleep before becoming beetles. One of my beetle-larvæ, having evidently attained its full size, seemed to be incapable of crawling out of the water. Hoping to save him I assisted him out on a pot of earth, and waited to see him bury himself. But evidently the dark and narrow tomb had no attractions for him. He refused to act as his own sexton, so I returned him to the water, where, in spite of the earth, with which he was furnished a few days after, and which he manipulated with his jaws, seeming to have a sort of cobwebby substance between them, he at last apparently drowned. Taking him out, I placed him on the earth again, but he was too far gone to do more than faintly wriggle his feet and tail, perchance, in token of adieu, for he died in truth this time and never had the pleasure of burying himself. Ah, well! "The good die young."

The full-grown, perfect beetles that come from these larvæ have a habit, when sitting still, of holding their long oar-like hind-legs curved up over their backs, instead of letting them lie stretched out in the water, the way the water-boatmen do with their oars. Another peculiarity of the *Dytiscidæ* is their bubbles. You should see a dozen dark-brown beetles, some of them perhaps an inch long, standing on their heads at the bottom of a jar of water, each beetle having at its posterior end a shining round bubble of air. Occasionally, from some collision or sudden calamity, one of the beetles looses his bubble. Up the round thing flies through the water to the surface, and the bubble-less beetle is seldom long in rushing up to protrude the end of his body and grasp another round bubble with which he comes rushing triumphantly back to his brethren. Even if, in the darkness of some obscure corner of the jar, the *Dytiscidæ* cannot be seen, yet one can catch the shine of their bubbles and know where the beetles are. The *Dytiscidæ* are of a very retiring disposition as long as they think that there is any one around. They are capable of concealing themselves pretty well. There may be a dozen of these beetles in a jar, and if there are only mud and weeds enough at the bottom, the creatures, when alarmed, will conceal themselves so that one would not know that there was a beetle in the water. The dark color of the beetles is easily concealed by its likeness to earth color, and, unless their bubbles betray them, the *Dytiscidæ* are safe.

There is a look of positive intelligence to many of these beetles. They are the Yankees of Water Land, in the matter of brains, though charity forbid that I should liken them in the matter of dispositions. When one of these beetles manages to slip out of his jar and dry himself in the warm sun, he will look at you in a knowing way as you go to catch him, and then, spreading his wings, will fly across the yard to some tree. It would never do to put such innocent-hearted beetles as the *Hydrophilidæ* in with the Water-tigers, unless one wished to see a battle in which the latter would come off victorious, the flesh-eater triumphing over the vegetable-eater.

The larvæ of the *Hydrophilidæ* and those of the Water-tigers are much alike, when both are half-grown, and the would-be "bug-sharp" will suffer many things and be often distracted by resemblances while learning to distinguish the one from the other. At times the indignant student will be ready to rashly affirm that the only way of telling the two apart is to wait till they transform to beetles, and if a *Hydrophilidæ* beetle comes forth, then the larva was of that family, and *vice versa*. And as, under the care of an amateur, most of the larvæ die before reaching maturity, any one can see that this experimental method of discovering the difference between the larvæ is not very satisfactory. One may know them apart, however, by the more clumsy form of the *Hydrophilidæ* larvæ, and by their dark, thick-looking skin and toothed mandibles. *Ver assassin*, the people of Europe call the larva of their *Hydrophilus piceus*, and certainly no one could tell the larvæ of the two families apart by reference to ferocity, for, if one family are assassins, so are the other.

In mid-April, as one drags the Water-tiger larvæ from their hidden nooks under the grass that dips into the stream, a flash of vivid yellow comes by, and one admiringly watches the first Colias butterfly of the season flit on black-bordered wings over the fields that are yellow as itself with the hosts of buttercups and the fewer but more brilliant blossoms of the California poppies. By the way there is a slight difference between the California Colias butterfly, as I have examined it, and the Eastern variety *C. Philodice*, as described by Harris, a difference that is interesting to those only who are concerned with the minutiæ of insect life. The difference is to be found on the under side of the hind wings, where the small secondary spot on each wing is not white, as in the Eastern variety, but rust-colored. Moreover this little spot is outside of the rust-colored ring that surrounds the larger white spot, instead of both spots being joined together and both surrounded by the rust-colored ring. Other mid-April visitors to the brook are little Agrion dragon-flies, and one may perhaps see the first of the large blue-and-black dragon-flies at the same time.

One year, when spring arrived, I was seized with a great desire to achieve that which I had not before, and carry some of the Water-tigers through their successive stages and observe their transformation into beetles. Accordingly I procured a collection of these larvæ from the brook. One of these larvæ was Oliver. He was so named because, like Twist of trite fame, he continually sighed for more polliwogs. The number of these interesting creatures that Oliver and his brethren devoured passes belief. My life was made a burden to me, owing to the necessity of frequent trips to the creek to get polliwogs enough to sustain the breath in life of Oliver and his brethren. I was also compelled to become a constructor of imitation ponds, lest the larvæ should want to burrow in the earth at times when I was not around to give them any to burrow in. Oliver's pond was constructed of a big, cracked marble basin, that had once been stationary but was so no longer, and a porcelain lid of capacious diameter. The lid being turned upside down and filled with water made Oliver's pond, and this being set in the middle of the basin full of earth, behold the lake was complete.

A section of mosquito-bar was tied over the whole so that should Oliver feel inclined to wander from the limits of the basin and make his tomb in the round, big earth, he could not do so. The search for his sepulchre would certainly have been discouraging then. But Oliver was capable of giving me any amount of discouragement.

After all my tribulations, early one morning toward the end of April, I found he was not in his lake. Joy filled my soul, for I knew he had buried himself. He had stuffed himself with the slain and had left two live polliwogs to mourn his departure.

Seizing a spoon I began to dig, feeling confident that I should find Oliver's tomb in a few moments, and intending, after looking at him, to cover him gently again and stick up a tombstone in the shape of a chip, and then patiently wait the two or three weeks necessary before I should hail the arisen Oliver in the guise of a beetle.

But I dug and dug and still no Oliver appeared. I think I can safely say that seldom since the days of my mud-pie infancy have my hands been so encased in mud as on the day of my dig after Oliver, I crumbled the wet earth till at last I reached the bottom of the basin, and yet I had found nothing of the one of whom I was in search. But, alas! I had found something else. It was some secret holes that I had forgotten, the holes that in stationary basins lead to the overflow pipe. If Oliver found those holes there can be no doubt that he managed to squeeze himself through and depart down the pipe to the outer world. In despair I replaced his lake, and tied on his mosquito-bar. I left his dish full of water, that if the wanderer returned he might find a place in which to swim, but my heart told me that I should never behold Oliver again. On mature consideration I came to the conclusion that Oliver slipped under the mosquito-bar someway instead of going down the overflow pipe.

About this time also perished the Scarer of Soap-Dish-Lid Lake. His name was given him because his pond was formed of a soap-dish lid turned wrong side up, and he scared every polliwog that visited him. The cause of the Scarer's death was the exuberance of Sol's rays.

Thinking my larvæ protected enough, I left them one hot day, but when I returned the Scarer could scare no more. The water of his lake had become too warm, and I barely rescued his brother, the Conqueror of Coffee-Pot-Lid Lake, from a similar fate. I might as well have let him alone, however, for although after a day of languishing he seemed to recover his health, yet he died within a week. I had removed all the larvæ's lakes to the shade of a high board fence where they might surely have been cool enough, but the Conqueror's first taste of warm water seemed to have been too much for him.

The Frightener of Flower-Pot Lake disappeared in the same mysterious manner that Oliver had done. I never saw such exasperating ingenuity as was displayed by some of my prisoners.

One day I came upon Conqueror II. of Coffee-Pot-Lid Lake outside of the water, lying on the earth. I supposed that he would dig his hole before my eyes, but after my coming he walked around a little and then plunged into his lake once more and hid beneath a sprig of alyssum, remarking that he was just as proud as some other people, and if there was any digging to be done he preferred to have no visitors during the performance. I told him that that seemed to be the opinion of all of his brethren, but he should remember the saying of Shakespeare : —

> "Oh, a pit of clay for to be made
> For such a guest is meet."

He made no answer to this bit of good advice, but he must have taken it to heart, for in the afternoon when I visited him and gently assisted him partly out of the lake, he took the hint and came entirely out. Poking his head down under a clod, he remained in that position quite a while, perhaps revolving in his mind the momentous question whether the earth was indeed a better place of residence than Coffee-Pot-Lid-Lake. He occasionally wriggled his tail, breathing yet through that useful member. He gradually became drier. He had been afflicted with quick, convulsive pants, presumably useful to him in getting rid of the water that clung to him. Half an hour went by. A mosquito, delighted with the stillness of the observer, persistently endeavored to make a meal off that much-enduring person, but was at last slapped to death. Various flies were also afflicted with curiosity as to why any mortal should be sitting on the piazza in their way. Growing impatient at last, I picked a sprig and gave the Conqueror a little prick therewith. So successful was this manœuvre that the surprised Conqueror executed a somersault and landed on his back. I think he had been asleep. Recovering his dignity the Conqueror turned his head toward his lake. But he remained on land. I was called away on an errand, and when I returned some two hours later the Conqueror was still in an undecided frame of mind.

Thinking that perhaps he expected me to perform the office of sexton, I poked a hole and assisted him into it. He seemed gratified, but the next day he came out. His tomb did not suit him, so I dug him a hole long enough for him to lie down in comfortably. This suited better, but it seemed to be a vexed question in Conqueror II.'s mind as to which was the more proper, to be buried with his head to the west or to the east. He lay facing the east a while, but afterwards changed to the other position.

I felt conscience-stricken. Perhaps I had unjustly accused the Conqueror. How did I know he could dig? What had he to dig with? His legs looked too weak, but I had thought he would use his mandibles, in the way that I had seen *Carabidæ* larvæ make holes in the earth. But the ground-beetle-larvæ had always lived in the ground and been used to the menial employment of digging, whereas the Conqueror had had nothing but water to go through.

It was the righteous larva of Yeast-Powder-Lid Lake that undeceived me. He was missing one morning, and digging after him with a heavy heart, — for I had learned to expect that my captives would escape, — to my joyful surprise I found him nearly at the bottom of the earth in the flower-pot. He was probably aiming at the hole in the bottom of the flower-pot, but I stopped that up immediately.

Having thus learned that these larvæ could dig, I went indignantly back to the Conqueror and dumped the earth in on that impostor. He was lying facing the west, but I gave him no time to assume a more orthodox position. A person that would tell as many lies as the Conqueror did has nothing to do with orthodoxy anyway. The Conqueror was like all the rest of us. He preferred to have some one else do his digging for him. The more I study bugs, the more do I perceive the resemblances between them and human beings. Nevertheless I did not relish being appointed Sexton of the Graveyard of Beetles.

My record of the Conqueror for the next few days is that he arose from his grave mornings, and was promptly buried again by the Sexton. I know not the number of times, but at last the Conqueror took up his residence under a tin lid and refused to stay buried at all.

The proper way for these larvæ to do is to make round cells for themselves in the ground. I dug down to see one of my larvæ that had been buried about a week and found him in quite a finely made cell that he had hollowed out. He was indignant at my intrusion, however, and I withdrew. He was the same larva of Yeast-Powder-Lid Lake that had informed me of the baseness of the Conqueror. After this larva had been buried nineteen days I again dug down to see how he was getting along. He had split a hole in the skin of his back, and four or five white segments of his body bulged out of the rent. He wriggled, and I again withdrew.

Will it be believed that not a single larva arose as a beetle? There was at first a slight doubt in the case of the larva of Yeast-Powder-Lid Lake, for I found his empty husk, but I know now that its emptiness was due, not to a secret resurrection, but to the ants. After all my labors in digging earth-worms and in going after polliwogs, this was my reward.

The following are the names of the unresurrected dead and of those that died before burying themselves: —

1. Conqueror of Coffee-Pot-Lid Lake.
2. Conqueror II.
3. The Hesitator.
4. Larva of Yeast-Powder-Lid Lake.
5. Scarer of Soap-Dish-Lid Lake.
6. Triumpher of Tin-Pan Lake.
7. Monarch of Mortar Lake.
8. "The Last."

List of those that "played hookey": —

1. Oliver.
2. Frightener of Flower-Pot Lake.
3. "The Last-But-One."

Numbers of other larvæ of no names perished in their infancy. The Monarch of Mortar Lake, however, deserves honorable mention. I am almost certain that, if the Sexton of the Graveyard of Beetles had left the Monarch alone, he would have come out as a beetle. But said Sexton was too officious, and to that may be ascribed the Monarch's death. I am sure his intentions were all right. He was the only one of all the larvæ that managed, after forming a round cell, to pull off his skin and become a pupa.

The Monarch had left the watery for the earthy element on the 7th of May. The 8th of June an unlucky fit of curiosity seized me and I explored the depths to see what had become of him. I found that he had made a good-sized round hole as big as a small potato. Inside the cell was the old skin overcoat that the Monarch had worn as a larva. Outside, below the cell, having evidently tumbled out through my fault, was the Monarch himself, but so changed that one would hardly have known him, for he was very much fatter than he had been as a larva, and was now a big, white pupa, with legs folded on his breast, and dull, black eyes showing under the white. He looked as if he were made of condensed milk. After this the Monarch turned dark, but the poor fellow never came out, for the Sexton had been his murderer.

The next time that I undertake to raise

Water-tigers will probably be when I can afford to hire a small boy to bring me a small pailful of black toad-polliwogs daily.

Dytiscus pupa.
The Monarch at rest.

I have read that in 1258 the commons of Castile bluntly required the king "to bring his appetite within a more reasonable compass." And the king meekly assented to the proposition. But what is a king compared with a Water-tiger? The former might agree to being deprived of anything he wanted to eat, but the latter would rebel. Pliny tells of men living on the smell of the river Ganges. But though the odor arising from some of the pools in which Water-tigers live is quite perceptible at times, yet the creatures hardly seem to thrive on that alone. For Water-tigers are not candidates for Nirvana, and therefore do not subsist on "insipid food," as those worthies were expected to. Like the Macdonalds of Glencoe, the Tigers have, indeed, "been guilty of many black murthers." But then, one can hardly expect the Water-tigers to practise such abstinence as the Druids were wont to use when they, in order to accustom themselves to curbing their appetites, would have a banquet prepared, and would survey the feast for some time. Then, their firmness having been sufficiently tried, they all withdrew without having eaten a morsel. Water-tigers never would think of such abstinence, and you shall find in their homes as many relics of dead and gone beings as the churches of Europe contained in the days when Geneva showed a piece of pumice as the brain of the apostle Peter, and a bone of a deer as the arm of St. Anthony, and when other places contained the hair of the Virgin, the tooth of John the Baptist, the shoulder-blade of Simeon, and a lip of one of the Innocents.

Perhaps, for the benefit of those who might wish to go dredging themselves, I should describe the implement with which I catch water-creatures. There is no need of spending a cent on apparatus for catching such insects. My dredger is of my own manufacture and consists of a strong, round, iron hoop that was probably once on a keg or something of the sort. To this hoop I have fastened a strainer consisting of a piece of an old calico apron. Occasionally the calico tears, but it is easily mended and is better than mosquito-bar because that will be likely to let small larvæ escape through the meshes. The handle of the dredger is an old round stick about a yard long. I think it was once part of a wooden clothes-horse. One end of the stick being split admits the iron hoop, and the two are firmly bound together by strips of stout cotton cloth.

Armed with such an implement as this, one can sweep under the water-weeds and be victorious. But use old calico or something thin for the strainer, or your patience may be exhausted while you wait for the water to run out so that you can see your captives.

Glass fruit-jars or old jelly-glasses make fine homes for beetles and bugs that do not need to come out on dry land at times. A stick or bunch of grass is all the resting-place needed by such insects.

Home-made Dredger.

The finest and cheapest receptacles that I have

ever found for caterpillars are empty yeast-powder or preserve cans. In some families it is a problem to know what to do with all the plum, peach, or tomato cans that accumulate. But when there is an amateur entomologist in the family, that difficulty vanishes. There is nothing so adapted to caterpillars, according to my experience. Tins are better than bottles. There is a warm, unhealthful atmosphere in a bottle that is not found in a tin can. Rinse the tins that no juice may remain from the former contents, for that will draw ants. One can keep caterpillars in such a tin till they are fully grown, and oftentimes they will go up and form their chrysalides on the mosquito-bar that is tied over the top to prevent the caterpillars from running away and to allow a good circulation of air. Lady-bug larvæ and other creatures will come to perfection in such tins, which should be put in some shady place where the sun will not heat them. One may learn the habits and observe the customs of all the creatures one can find in the neighborhood, without spending anything unless it is for shoe-leather.

But do hide your tins and yourself when you are at such work, or you will hear a window open, or else it will open without your hearing it, and you will have to endure some such talk as this: —

"What are you petting there?"

"Caterpillars." I stand up and suffer a faint wonder to pass through my mind if I am never to be safe from intrusion, even in my own yard.

"What do you do with them?"

"Keep them till they turn to butterflies."

"What do you do with them then?"

"Let them go."

"Oh, you just keep them to " —

"To see them transform."

"What?"

"To see them transform."

Amazement evidently filled the soul of the neighbor.

"I never saw them," she said.

And this from a woman forty-five or so, with eyes in her head. What sort of a world do such people live in, that they never notice anything? And what, I wonder, can be that woman's idea of some of the verses in the Psalms? I presume she has read them a hundred times. "Praise the Lord from the earth" ye "creeping things," says David. What are the "creeping things," according to this woman's eyes? I wonder if she ever explained that verse to herself.

I remember the day when the idea first entered my brain that other creatures than human have interesting lives. I must have been about eight or nine years old. I had been taking a walk in a little mining town among the Sierra Nevada foothills. A minister was with me, and he had a hand-microscope or glass in his pocket.

We sat down under the trees on a hill somewhere back of a church, and he showed me through the glass a multitude of little creatures living in the heart of a yellow flower. And he said to me that perhaps those tiny folk had houses and cities inside that flower, just as we bigger people had outside of it. It was a wonderful thing to me to look at the mites and imagine what the minister suggested, and I have a fancy that perhaps I longed to be one of those little things for a time, just to see the supposed houses and cities. But I remember that as a very wonderful day, one on which I did not see half enough through that glass to suit me, but still one on which I obtained a glimpse of a world very different from that in which I usually lived.

No wonder people unacquainted with such a world ask questions. Some day I may write " A Catechism for The Warning of Those Who Propose to be Bug-Hunters." It will contain a number of questions that they may expect to be asked and that they must always be prepared to answer, if they do not wish to be thought idiots. It is quite probable that they will be thought such, anyway. Some of the questions will be these: " What are you doing?" " What are those?" " What are they good for?" " What do you do with them?" " What are you after?" " What do you get them for?" " What makes you like them?"

This is a fearfully inquisitive world.

But there is one question that is seldom asked of a bug-hunter. It is this, "Who made them?" The majority of people scarcely pause to realize that the different kinds of creatures represent so many of God's different thoughts, and that it might possibly be worth while to glance at the things that He has deigned to place on earth.

CHAPTER VI.

WHIRLIGIGS.

"Here's another ballad of a fish."
Winter's Tale.

THE end of May has come. Filaree, with all its spikes, is abroad in the land. Mustard shines like the sun. Blue-grass blooms. So do blue-bells,—*Brodiæa*. White Yarrow shows his snowy head beside the stream. Blue lupines are mixed with yellow poppies among the already browning grass by the roads. In one place by the brook is a big clump of sticky *Mimulus*,— *M. glutinosus*. The locust-trees are full of perfume. A great yellow-and-black *Papilio*, swallow-tail butterfly, flits above the grain on the hill-top. Many wicked little green beetles, *Diabrotica*, the pests of fruit-growers, crawl over the

Blue-bells.
Bud and flower of *Brodiæa terrestris*.

stems of the blue lupine that is just now in its glory.

Mimulus glutinosus.

Let us climb this bank. Kind Neighbor Thistle, give us a hand. Benevolent Brother Teasel, do likewise. It is only in that one place that the *Mimulus* grows. Perhaps it may die out from this region altogether some day. Wild flowers do die out. I can remember when shooting-stars, *Dodecatheon Meadia*, cowslips, grew near here, but it has been

Shooting-star.
Dodecatheon Meadia.

many a day since I gathered them in this district. There are none here now, and I hope the same fate will not overtake the butterfly lily that comes in but scanty numbers among the dry grass of these hills in June.

From this height we may look down and see the Whirligig beetles on the surface of the pools.

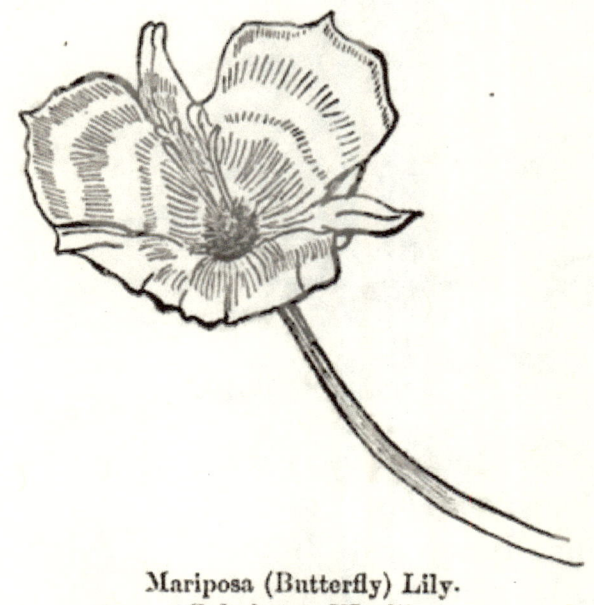

Mariposa (Butterfly) Lily.
Calochortus Weedii.

"Death's net, whom none resist," is often more full of Whirligigs than is the dredger of the persistent, unsuccessful mortal. A dead Whirligig beetle is a solemn sight. The one that was the liveliest beetle of the brook, an animated flash, a perpetual whirl, the one that seemed as if he could not move quickly enough to suit himself, lies in your hand, stiff, with his legs folded in death. But if the beetle himself is solemn, sometimes his eyes are ghastly. The compound eyes of these Whirligigs are very curious objects. One would surely think in looking at the dead beetle that he had four distinct eyes, two in the usual place, and two others under his chin, so to speak. In one beetle, long-dead, that I examined, the under eyes were the more ghastly. They looked white, like a human eye that had neither iris nor pupil, while the pair of eyes above showed black inside a rim of the same whitish color. In a beetle just dead both pair of eyes look black. The eyes under the chin, of course, are the ones that the Whirligig always keeps under water to see if any enemy is coming up from below.

How can a beetle be sensitive to knocks when he is incased in such an armor as this? The *Gyrinidæ* look as if they had suits of steel. Not for them are the scratches and bumps of life. It is enough to make human beings wonder how it would seem to be so safely shut in from all pricks, and one looks at one's finger-nails, remembering the old Jewish tradition that Adam and Eve at first were entirely covered with finger-nail, but, after the fall, this invulnerable panoply dropped off, and mortals since then have had skins that could be hurt. Nevertheless the finger-nails were left that the first pair might always remember Eden's freedom from pain, when they looked at their hands.

If, as in Adam and Eve's case, panoply against harm argues innocence, then these *Gyrinidæ* might claim to be the most perfect of beetles. But alas! here is another case of duplicity in this brook, for the *Gyrinidæ* are black murderers, killing by trade, and when one sees a host of them whirling over some corner of a pond, it is as though one were looking at a band of pirates, for every available small insect that comes within reach of those *Gyrinidæ* is doomed. Sociability is a characteristic of the *Gyrinidæ*. It is seldom that a single one goes skimming by himself. Generally they prefer to whirl in small crowds. Beholding their evil deeds one might think that the *Gyrinidæ* were sociable in order to keep each other in countenance, much as a band of sworn cut-throats might be.

Still, during the last few days of February or the first of March, before the *Gyrinidæ* have come out in great numbers, there will here and there be found one whirling by himself. It requires some skill to catch a beetle of this sort, unless one has learned the trick of confusing him by splashing with the dredger. The beetle will whirl and whirl till one's eyes are blinded and one's head is dizzy with the effort of following the motions; and then, unless a fortunate scoop is made suddenly with the dredger, down goes the beetle under the surface and is lost to view. Moreover it is a wise person that can keep a

Whirligig after catching one. As a youth once said to me, after looking at my bottle of captured Whirligigs, "They act as if they were crazy." Moreover, Whirligigs can climb glass, and if one takes off the mosquito-bar carelessly from the bottle, as likely as not a Whirligig drops over the side. It is a good thing that these *Gyrinidæ* cannot walk well. If they could walk out of water as well as they can dash around on top of that element, there would be no catching them. As it is, however, they are easily caught on dry land, for they have to move at a pace that must be exasperatingly slow to them, since their feet are not suited to crawling. Then it is that one may observe the relative difference in the pairs of legs of these *Gyrinidæ;* the first pair being the longest and capable of being suddenly thrust out at the prey, and the hind pair being flat like oars, but like oars with short handles, not long ones, as in the Water-Tigers. The *Dytiscidæ* have long hind legs, the *Gyrinidæ* long fore ones.

"They call them 'good-luck bugs,'" said a little fellow to me once at the brook, as he looked upon the Whirligigs.

"What do they call them that for?" asked I, in search of information.

"Because they bring good luck," answered the boy, confidently.

"Do they?"

"Yes 'm. You just take and keep one of them and you'll have good luck."

"Why, is n't that strange! Did you ever try it?" asked I, being bent on finding out what the little revealer of superstition would say.

"Yes 'm," responded he; "I kept some of them, and I always had good luck, every day 'most. One day I found two bits;" and, having proved his theory to his own satisfaction, and demonstrated his right to be called a Californian by his use of the common expression for twenty-five cents, he proceeded to assist in the dredging, innocent of the knowledge that to be a healthy, dirty boy, with no care but to hunt red-legs in the creek, is indeed to "have good-luck every day," without any assistance from the virtuous Whirligigs.

But if he was mistaken in regard to the blessedness of possessing these beetles, this boy had observed their habits, for he informed me that when he kept the "good-luck bugs," he gave them "about five flies" every day at noon, and that the Whirligigs would jump for them, as soon as they were thrown in, and would eat them. This statement can be verified by any one who is willing to catch a fly for Gyrinus. Three or four beetles will try to devour a single fly; in fact, on an especially hungry day I have seen eleven Whirligigs form a circle around a fly and grasp it, while on the outside of the circle were still others trying to poke their way in to the centre of attraction. I do hope that boy killed the flies before giving them to the beetles. Had I known the habits of Whirligigs as well then as I do now I should have entreated him to be sure that the flies were dead. Not that I am an admirer of the fly. On the contrary, I abhor him. But one does not like to see him killed by particles.

Just about the time when Whirligigs first begin to be plentiful in the spring, the last of March, there will come up in one's dredger what look like bits of dry tree-twigs. They are five eighths of an inch long, or thereabouts, composed of eight segments looking like the divisions of a twig. But put these sticks into water, and

Pupæ of *Tipulidæ*.

they will float perpendicularly, not horizontally after the manner of real sticks, and one will notice that from the part that is uppermost there stand out two little projections. Some day when you are looking at these "twigs" one of them will give a kick, and the next morning, when you go out to look at the bottle, you will find the twig with a hole in the upper part of it, and in the upper portion of the bottle, holding itself above water-line, will be a long-legged fly. This fly came out of the "twig," for that was the brown case, covering the pupa of one of the flies that look like giant mosquitoes, the *Tipulidæ*.

Daddy Long-Legs the children call the fly related to this that passes the two earlier stages of its life in the ground. "Crane-flies" others name them, and the French cry "Tailleurs," or "Couturières." I think these water-flies sometimes drown while struggling out of their pupæ-cases. At least I have found one in such a position as to convey that idea, and the wonder is that any of them get out alive when there is nothing near to be grasped by the insect during the operation.

If you know where to look you may find other Tipulid larvæ near here. A few live-oak-trees are scattered

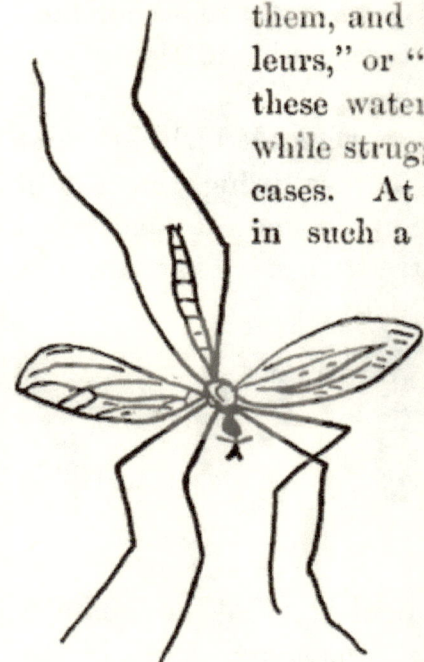

Tipulid Fly, — somewhat larger than mine.

along the banks. Some of the trees are old, for holes in their trunks show the rotting wood within. One day as I wandered here I spied in a tree a hole about on a level with my head. The hole extended through the tree so that I could view the interior. Black-bottomed with the débris of tree-decay, the hole looked mysterious. Should I explore it? I did. My trowel came forth with its burden of black. Spreading it on the ground I looked for live inhabitants in every trowelful. A mosquito walked sleepily out of the débris as though deprecating my actions. A small centipede, I believe, threatened me with destruction. But in the mass I found the larvæ of ten Tipulid flies, one wedged into the side of a piece of wood so soft that I could mash it between my fingers.

The larvæ were light-lilac-colored worms from seven eighths to an inch long, with bodies ending in a round of V-shaped points, somewhat like some of the larvæ of dragon-flies. The Tipulid worms had an uncanny way of moving, the black internal arrangements showing in places under the almost transparent skin, giving the observer the impression that the creatures might be breaking to pieces. Extremely uncomfortable were the larvæ on being dug out of their tree-home, and unanimous were the ten in their appreciation of a jelly-glass of the tree-mould in which they might hide again. They soon turned to pupæ, for they were nearly grown. Such larvæ live upon vegetable mould. Réamur found that such creatures extract all the nourishing matter from the earth that they eat. One would think it a pretty dry diet, yet the larvæ seemed to enjoy life in the old tree that opened its trunk and said to the insect creation, "Come, little friends, I can help you still. No matter if I am old. I can help you better on that very account. Whereas, before, the most of you had to stay outside, I can take you now into my very heart."

All who have kept Whirligigs know that they, like the *Dytiscidæ*, carry down bubbles, and you may see a beetle at the bottom of the jar, holding to some little pebble and accompanied by his tiny quota of air.

Figuier tells us that in a little lake of Solazies, Reunion Island, there are some tropical *Gyrinidæ* of a somewhat large size, and that the patients, who go to Solazies for the mineral waters, amuse themselves there by fishing for the Whirligigs with lines baited with bits of red cloth which the beetles will attack. Some of my little Whirligigs are also deluded by a snare of this kind. Fishing for them with a bit of red braid attached to a red thread, one will see some beetle seize it and hold on till he is brought almost out of the jar, while he is vigorously seeking for nutriment apparently in the braid. A white bit of cloth has also some attractions. I suppose it reminds the beetles, at first sight, of a white moth. Black cloth seems to have almost no interest for the Whirligigs.

The bill of fare for these *Gyrinidæ*, as far as I have observed, is as follows: —

1. Middle-sized Spiders. (Often taken alive, on the whirl.)
2. Dead Flies.
3. Live black gnat-worms.
4. Live Aphides.
5. Dead small Moths.
6. Dead Daddy Long-Legs.
7. Dead Mosquitoes.

I think that Whirligigs are cannibals, since I have found them floating dead in the jar with their heads off. Still, I never saw a Whirligig kill another, but appearances are against them.

Whirligigs seem to regard ants as unworthy partners in this dance of death and discard them. I should not think myself that an ant would be very good eating.

I do not remember ever having seen a jumping-spider discomfited by these beetles. The Whirligigs will skim by and longingly nip at a jumping-spider's toes, but evidently there is considerable fear mingled with the admiration, even though the beetles must see that the spider is in an element to which he is unused.

Smaller varieties of spiders are speedily disposed of. I was once present at a duel between one of these spiders and a Whirligig, in which I thought at first that the black-mailed warrior would be forced to yield, for the spider waved his legs over the Whirligig's head and evidently was quite anxious to bite him. But the Whirligig, undaunted, plunged into the spider's embrace, and, before I could divine his intention, that Whirligig bit his antagonist directly in two at the waist and whirled off with the fore part of the body, leaving the hinder half to the care of some less brave Whirligigs. So ended this " Gentle and Joyous Passage of Arms."

Beetles of other sorts seem to be rejected by Whirligigs, according to my observations. Probably the hard outside of most beetles is what deters the Whirligigs from attack. A dead bee that I once gave them was merely tasted of. Bee-flesh and ant-flesh are alike distasteful, which shows that Whirligigs have some gustatory powers, as well as other folks.

" *Tourniquets*," turn-stiles, or turn-pikes, the French call these Whirligigs, and, if one of these beetles were set for that office, I think he would whirl as rapidly as did ever any turn-stile. Contrary to what one would think, these lively beetles seem to be quite well contented as captives. A jar is a world almost big enough for a Whirligig.

" Beetles," says old De Mouffet, " serve divers uses, for they both profit our mindes, and they cure some infirmities of our bodies." I do not know that Whirligigs are remedies for any of the ills that flesh is heir to. If, then, these beetles serve no use for our " infirmities," how do they " profit our mindes " ?

Well, perhaps they may remind us of the old,

old truth that, be we nimble as we may, Death is quicker still, and shall sometime overtake us. And then, if we have spent our lives snatching good things away from our neighbors, and counting all as our enemies, only valuing others for what we can take from them, we shall not be missed or mourned for, though we no longer glitter in the sun. Nature itself teaches the lesson of death; and Louis XIV., used as he was to adulation, must have despised Bourdaloue who, having cried out in a sermon, "All, all must die!" seeing the king start, hastened to reassure the royal mind by adding, "*Almost* all, Sire!"

Death is the swiftest of all things. Is that the lesson of the Whirligig, living and dead? Perhaps some observer, more wise than I, can find something in the *Gyrinidæ* that will more profit his "minde" than this lesson of mine. To every one his own perception and the responsibility thereof. Only it does no good to learn a lesson from a book or from nature, unless the learner straightway puts it in practice; and a warning of an event is of no value unless it leads those warned to make ready for it.

The pictures of water-beetles perhaps give some people too good an impression.

"I'm mamma's little water-beetle," said a golden-haired little two-year-old to a friend of mine who had been showing her the pictures of insects. Perhaps when the child is older, and when she better understands the murderous character of many water-beetles, she will not be so ready to name herself after them. For the carnivorous water-beetles are as bloodthirsty as were the mob that Machiavelli tells us tore poor Ser Nuto to pieces. Says that chronicler: "Ser Nuto being brought by the mob into the court, was suspended from the gallows by one foot; and those around having torn him to pieces, in little more than a moment nothing remained of him but the foot by which he had been tied."

Scarcely so large a remnant as that is left of some victims after the water-beetles have done with their prey.

The willows border all the stream here for a short distance. They spring up at irregular intervals in other portions. In early June one may walk among these trees and note the many galls that swell out from the willow leaves. The galls are about as big as cherry-stones, or bigger, and there is usually one gall to a leaf, though two are not uncommon, and I have found a leaf with four. Open one of these galls and you will find a very small, yellowish-white worm with six legs and a brown head. Look at the head under a microscope, and the worm appears exactly as if he had clapped on a brown night-cap. This night-cap is hairy, and two black eyes stick out of it. It is very hard work for the worm to walk on the glass slide, and one can have a good look at him in the midst of his contortions.

I wonder if such willow-galls have never been used in augury, as have those of the oak. Gerard in his old " Herbal " tells us that " the oke-apples being broken in sunder about the time of their withering doe foreshow the sequell of the yeare, as the expert Kentish husbandmen have observed by the living things found in them ; as, if they finde an ant, they foretell plenty of graine to ensue ; if a white worm, like a gentill or a magot, then they prognosticate murrain of beasts and cattell ; if a spider, then (say they), we shall have a pestilence, or some such like sickenesse to follow amongst men. These things the learned also have observed and noted ; for Matthiolus, writing upon Dioscorides, saith that, before they have an hole through them, they containe in them either a flie, a spider, or a worme : if a flie, then warre insueth ; if a creeping worme, then scarcitie of victuals ; if a running spider, then followeth great sickenesse or mortalitie."

On these willows, too, in June, I once found a big, green, Sphinx caterpillar, the same shade as the willows, just about a match. He had a spike of a tail, after the manner of his race, and that tail was gorgeous, being blue above and red beneath. He had eight little buttonholes of pink spiracles on his sides, and was dotted all over with fine white points. There was a well-developed stripe of yellowish-white running obliquely down each side from the tail, and five other, scarcely noticeable, small stripes ran parallel to this on each side. His ridiculous-looking head was bordered on two sides by a rim of yellow, like a picture set in a brass frame, so that his head had the shape of the letter A, without the cross-mark. Besides this he had two white lines that ran the length of his sides. Altogether I was very proud of him, and was always getting scared over his supposed loss, he being so much the color of the willow-leaves that I was always missing him, even when before my eyes. At last I learned to look for his tail and then I could find him. But alas! one day I could not find his tail. Neither did I find him. The sad truth was apparent. Sphinx had run away. I had not thought such baseness possible in so solemn a creature, always standing in one position, with his head in the air, deep in meditation. But I never saw him again alive. However, several days after his disappearance, I picked up a jar of water that contained some pond-snails. The inside of the jar was coated with green, and I did not at first notice something of the same color lying at the bottom of the water. But I soon saw it. It was poor Sphinx. He had tumbled into the water and drowned. I helped his corpse out of the water but he never revived.

What a disreputable set some of these flowers along the fields are! For instance, take the buttercups. Have not **they reason to** hang their heads in shame when they remember how nearly they are related to **the** clematis that **the** English call Beggar's **Herb. This plant has leaves that, if** applied long to the **skin, will make sores, and beggars** are said sometimes **to use the** leaves for this purpose in order **to draw** forth people's **compassion** and money.

And Thistle **over** there is no **better, for do not** his folk make **such " thistleries "** in Paraguay that robbers **can hide** among them **and** attack unwary travellers? And was **not Thistle** anciently sacred to that disreputable **heathen god, Thor ?** Did not thistle-blossoms get their color **from the** lightning? I am afraid Thistle **is hardly respectable.**

And **as** for Mustard, **his very** name **is his disgrace, for does it** not show that **the ancient Romans** mixed mustard with their **sweet wine, or** "mustum." I fear that **Mustard has** been present **at many** an orgy. Such disgrace **is hardly atoned** for by the fact that the ground mustard that Mrs. Clements sent **King George I.** pleased that grumpy soul and caused his English subjects to approve of the yellow stuff.

White Yarrow, there, **is** hardly more decent, for are **not his folks used in** Sweden **in** making beer? And if you come to his other name, *Achillea*, did not Achilles kill Memnon, and does not said Memnon's mother, **Eos, weep** for her **son** every morning, **and so** form the dew? **If** you do not believe that, come here some morning, and look in the grass below *Achillea* and see if the tears of Eos do not lie there, at the murderer's feet. When I reflect on all the disgraceful histories of these flowers and their relatives I hardly feel like noticing them as I pass by.

As for the blue lupines that adorn the sides of the road on the hill yonder, they remind one of the dreadful mistake said to have been made by the Germans at one time. For these lupines belong to the *Leguminosæ*, the same family that contains the "sweet" and "everlasting" peas that blossom in pink and white and blue in our gardens. And it is written, — though whether in tradition or in history I know not, — that some Germans at one time thought that sweet-pea seeds would make good eating. So the seeds were ground and mixed with flour, and indeed fine bread was made from it. The Germans thought that they had discovered a wonderful way of proceeding, but after a while those who continually ate the sweet-pea bread began to find their limbs and joints becoming mysteriously stiff. They grew worse and worse, and, by and by, the poisonous bread made the people cripples for life.

Nor was this all. Some people had fed pigs on such meal, and it is said that the pigs, too, lost the use of their limbs entirely, and fell flat on the ground. Here was a sad state of affairs. Experiments are dangerous things. So thought the German government, and it sent out an order that no one should use any more of the poisonous sweet-pea bread. Such are the associations connected with the lupines.

Still, there are those less depraved among these flowers, and the blackberries by the cliff may remind us that their relatives, the strawberries, have comforted many a bereaved German mother in olden times. For when the people of Germany were heathen and worshipped the goddess Frigga, there was a belief among her devotees that on one day of each year Frigga, the invisible, went strawberrying, and when, afterward, she left the earth with her rosy load, she divided her berries among all the little children that had ever died. And, so firmly was this believed in Germany, that on Frigga's strawberrying day, no mother whose little child was dead would ever eat any strawberries, for, if she did, her little child would not receive any when Frigga divided her berries among the children in Paradise.

It is a heathen superstition, and yet who can tell how many mothers' hearts may have been comforted at the thought that by denying themselves they could add a little to the happiness of those they so sorely missed? And perhaps there are some of us nowadays who might be benefited by the truth at the heart of the old superstition. For there is a truth here, and it is this. There are sacrifices that we may still make for our holy dead. We may take up their work upon the earth. And to every one who does this comes the comfort of the knowledge that heaven and earth are not so far apart, after all.

CHAPTER VII.

WATER-LIZARDS AND THEIR ILK.

"Nay, good my lord, be not afraid."
King Richard III.

"There are lots of Water-lizards over in the cañon in the other creek. They're red. I go over there sometimes, and may be I'll bring you some."

This is the substance of a generous-sounding speech made to me by a lad, but his promise was never fulfilled. The brook and I heard it, but the brook was better than the boy. Either before or after this, I drew my first Water-lizard from these shallows.

A yellowish beastie he was, about two and three fourths inches long, with a black line running down the middle length of each side. His tail was flattened and spotted with black. He had both anterior and posterior pairs of feet, the forward pair being four-toed, and the hinder pair five-toed. Dark eyes were his, and there were three pairs of reddish-yellow gills. These were very noticeable, standing up like feathers on either side of his head just in front of his legs. These gills gave him a very peculiar look, reminding one somewhat of the ruffs that ladies wore in olden times. The pair of gills nearest his head stood upright or nearly so.

I tried to suit the poor fellow in regard to food, for I put into his jar a few water-shrimps, a small dragon-fly larva, a little red water-worm, an earth-worm, and a small scorpion-bug, but, although I do not know what better fare he could have found under the weeds from which I took him, my viands were all thrown away on him. He would have none of them. I think he ate nothing while he lived with me. He seemed to be very much alarmed whenever I came to visit him, and would race around the jar in terror, or look at me through the glass with suspicious eyes.

And then, alas! after he had sojourned with me for nine days, I came to his jar to find him lying on his back. He was dead, poor prisoner, and I kept him a long time afterward in a bottle of alcohol. But I regretted that I had not given him a larger dish and a place where he could have come out of the water if he so desired.

What is there about the lizard-shape that gives one a feeling of dislike? The creatures are often pretty in coloring, intelligent of eye, and yet one shrinks from them, whether in water or on land, whether of the salamander or of the lizard family. The matter of likes and dislikes is a curious one. Frogs are not so very far removed from the lizard shape, and yet with how much more of respect do we look at a frog. And a fat old grandfatherly toad that lived in a hole in our garden once commanded my childish interest, almost affection. But a lizard! No matter if Wallace does tell us that, on the Ké Islands, he found swarms of little green lizards with tails of the "most heavenly blue," no amount of coloring can make a lizard very acceptable.

On the hill away beyond those eucalyptus-trees, a party of us one May Day had an adventure with a lizard. The creature had a very long tail. One of the boys tried, with a lad's usual kindness, to stamp on the lizard, when, lo! it threw off the tail and ran for dear life.

But the tail! It went rushing around, looking like a little snake. Perhaps the tail was trying to run after the lizard. At all events our party left looking after the lizard to see what his lively remnant would do. No one dared touch the wriggling thing, until a sturdy carpenter snatched it up. He bequeathed it to me, and I put it in my handkerchief, but before I had walked the half mile home the lizard's memento was quite still. Having a number of Whirligig beetles at the time, I gave them an opportunity to taste of lizard's flesh, but they would have nothing to do with it. Whirligigs have likes and dislikes, after the fashion of human beings.

The Water-lizards described as living in the

other brook haunted me for many a day, till, at last, going forth once in April, a friend and I made the pilgrimage of several miles and received our due reward. I would advise all goers after Water-lizards to avoid taking a dog with them, especially a frisky dog. Such a one, black, with the merriest eyes I ever saw on a dog, met us and insisted on tagging. He was certain we were going to do something that he was interested in, and he invited himself with as much assurance as one would think those old Grecian dogs might have attained to in the days when polished Hellenes, sending notes of invitation to their friends, were wont to be courteous to the friends' dogs, also, and request that they might be brought along.

Triton.

This dog had been taking a bath in some pool, as his coat showed, and so enamored was he of the water still, that he rushed at the little wayside stream, plunging in with a vigor that would have made dredging useless, if it had been attempted. He came once to pat his paw down where I was striving to find some inhabitant in the mud.

Fortunately that dog was left behind before the brook was reached, otherwise the Water-lizards might not have been caught. The water under a bridge was quite transparent, and on turning over stones, two of the creatures came in sight. No more were to be found. Probably these had taken a morning stroll among those stones, and so were caught. Gentle creatures they seemed to be, not much frightened even when the dredger scooped them up singly as they were found, and they came out of the water, the drops falling from the yellow and gray of their skins.

Before leaving the brook, on vines or weeds beside it I found a number of the pupæ of Frog-hoppers. Spots of white froth looking like soap-suds were all that could be seen, but penetrating a white mass I found a little insect looking somewhat like a small lady-bug, the forward part of the body being black and white, and the abdomen red. *Crachat de Coucou* the French peasants call such spots of froth, and in England they are known as Cuckoo's spittle, or *Ecume Printanière*, — spring froth. That the cuckoo should be credited with such an overflow of saliva is a mystery only equalled by the fact that other credulous people assert the toad to be the owner of the " spittle."

One of my larvæ of Frog-hopper, — enlarged.

The Frog-hopper larvæ that I took home were of varying sizes, one being about an eighth of an inch long, others a little larger. In my zeal for Water-lizards I neglected these larvæ till the spittle of most of them was gone. Then I endeavored to give the creatures a new supply of fluid by putting cuttings of plants into their bottle. But honeysuckle and rose, chick-weed and lily slips had not the right taste. No weed I could find suited them, the kind on which I found them not growing here, and one by one my Frog-hoppers miserably perished, without having been able to produce any more froth. "The larva of the *Aphrophora* cannot live long out of its frothy envelope," says Figuier. My last one died five days I believe after I picked the weeds the larvæ were on, but the dabs of froth lasted during the first day or two, so that the larvæ were not dry all of that time.

However, on another day, beside that brook I found a mass of foam on a blackberry shoot, and, breaking it off, brought it home. The Frog-hopper larva inside that mass proved to be larger than any of those I had previously found. He was dark, almost black, with a few light marks.

Calling to mind De Geer's experiment with a similar larva which he compelled to make new froth, I resolved to imitate him. I drew my larva out of his bubbly world and tried to wipe him dry. De Geer thought that the froth serves to protect such creatures from the heat of the sun and from attacks by spiders and other carnivorous creatures. The froth serves, too, I think, as a sort of drowning-place for other little insects, as I found a small winged creature, perhaps a winged aphis, dead in the froth.

I had obtained for my larva a shoot of what I supposed was a cultivated blackberry, for I thought that he would not know the difference between the taste of that and the taste of the wild variety. My supposed blackberry shoot, however, was finally discovered to be a raspberry one. After I had wiped the froth from him as well as I could, so that, while not being exactly dry, he had not much moisture on him, he tumbled into the cup of water in which I had placed the shoot to keep it fresh. He descended to the bottom of the cup, but my rescuing finger was after him, and he clutched it and was saved. However, I did not wipe him dry after his involuntary plunge.

I put him on the shoot and he speedily began work. With the constant bending of the hinder portion of the abdomen, little bubble after little bubble collected under him. Within nine or ten minutes he had quite a number, enough to make a small mountain of froth. Still the whole upper surface of his body was uncovered.

In twenty-five minutes from the time of starting, the froth had mounted so high that it began to touch his back. Some of the time he kept his head down to the " blackberry " shoot, as though he might be drawing in juice, as these creatures do through their beaks. At other times he raised his head above the shoot, but the hinder part of his body was continually elevated above the fore part, so as to give him the appearance of being just ready to turn a somersault.

I thought that he was succeeding finely, but, about three quarters of an hour after he first entered into the business of making this batch of foam, the Frog-hopper larva left the place altogether and wandered to the end of the shoot, where were some leaves. He still retained some moisture on the under part of his body, but why should he waste the bubbles he had been making? There they were, a pile of froth, waiting for him while he crawled over the leaves. I picked him up and put him back in his place, but he would not stay there. Away he went toward the leaves again.

I put him back a second time, and again he fell into the water. This second bath sobered him, I think, for he recommenced work. Perhaps the reason why he left his froth was that he remembered that he had not explored the shoot, and, inasmuch as he did not expect to make any more journeys after the froth had once closed over his head, he thought he would stop work and travel a little. This was what I thought at first. But after he had made another pile of froth about as big as the former one, he again left it and wandered off.

The truth dawned upon me. That fellow was smarter than I had thought. He *did* know the difference between the cultivated raspberry and the wild kind of blackberry. He did not like the raspberry. Hoping that he would not oblige me to journey to the brook for his food, I gave him a shoot of wild blackberry that I had kept in a pail for the needs of any of my menagerie.

Frog-hopper did not like the looks of my present. He had never been taught the polite truth embodied in the maxim that one should not look a gift-horse in the mouth. That part of Frog-hopper's education had been neglected. He looked over my shoot but did not offer to make any froth. It was quite apparent that the shoot was not fresh enough to suit him, and he was waiting to have a better one appear. Overawed by his wisdom in regard to blackberry shoots, I put on my hat, snatched the scissors, hastened to the creek, swung myself under a fence, and, in spite of the proximity of a number of boys, secured my fresh wild-blackberry shoots, and came home.

That was exactly what Frog-hopper wanted, and, after considerable delay, he proceeded to bury himself in foam, and succeeded so well that at about half-past nine P. M., when I gave him a farewell look for the night, all that could be distinctly seen of him was a little black dot, a portion of the hinder end of the body. All the rest was covered in the foam.

Still it would manifestly be impossible to bring up such a larva on a shoot. This was shown next day when the branch, although propped by pieces of coal in the water, would not stand up securely, and Frog-hopper's mass of foam hung down so that it would not cover his back. He became disgusted and again went on his travels. So I journeyed to the brook and dug up a couple of scrawny little blackberries, planted them at home, conveyed Frog-hopper to the spot, put him on a leaf, and tied a cloth around that branch to make sure that I should see him again. I furthermore tasted both the cultivated raspberry and the wild blackberry, and I came to the conclusion that there is a difference in the flavor of the sap. The wild blackberry is more pleasant. I did not wonder that Frog-hopper knew that I had not given him the right thing at first. There is not much use trying to fool a bug. He is generally smarter than he looks.

I untied the cloth next day. Frog-hopper was there, but he was without any froth. Disgusted with his tribe, I bundled him into a tin, took him to the brook, put him on a leaf of a blackberry vine, and gave him my parting blessing. Such bugs are nuisances.

"Flea-grasshoppers," *Sauterelles-Puces*, does Swammerdam call these Frog-hoppers, because the adults jump like fleas. And one European kind, prone to live on fern-stalks and thistles, has the complimentary title of "*le Petit* Diable." But I fear these Frog-hoppers are to be classed among the injurious insects. It is very well for them to attack weeds by a brook, but the Frog-hoppers need not pretend that they never visit anything else. And many such creatures on cultivated plants would kill them, as the people of the Basses-Alpes could testify. For does not a Frog-hopper, *Jassus devastatans*, make bold to hop on their young corn and lay waste their cereals? Frog-hoppers are certainly bad creatures, not worthy of our further consideration. Let us turn from them to the subject of this chapter.

The first morning after I caught the Water-lizards, I thought one had escaped, for I went to the dish and found but one in the water. Further search, however, discovered Number Two snugly hidden in the cloth that I had tied around the dish the night before. He received the reward of his wanderings in being obliged to hang by one of his hind feet till I could cut a thread of the cloth that had become wound around one of his legs. His tiny little hand looked almost human as I held him during the operation.

This adventure did not deter him from going again and again into the cloth. I suppose he slept there sometimes nights. It became my custom to run my hand around the dish mornings before untying the cloth, and I often felt his body lying somewhere beneath. It gave me the sensation of a corpse under a grave-cloth. He did not attempt to run, but would wait for me to untie the cloth and take him up ignominiously by the tail and put him in the water. I have an idea that sometimes I put him in sooner than he liked.

The other Water-lizard was not quite so bold. I took her to be a lady. She was given to hiding under the stones I put in the dish, and but seldom did she ascend to hide in the cloth. Occasionally, however, I found both there, and put them back in their watery home.

Darby and Joan were the names of my friends. It was easy enough to tell when Darby had been spending some time out of water. His coat above would be of a dark color that would last nearly all day, even when he stayed in the water, and it would be late in the afternoon before he would be of the same color as Joan, a sort of grayish yellow. Both were brighter yellow beneath, and neither had gills. Their eyes had a greenish color, and the creatures were four-toed in front and five-toed behind. Their heads were much like those of frogs, noticeably so as I would sit looking at the dish and see Darby's head come rising up to the margin. Seeing no body, one would certainly have said that a frog was coming. Darby was six and a half inches long, for I measured him one day. It was easy enough to control him, and hold him in place by his tail while measuring him. I do not think that he admired the performance, although he never seemed to dare to refuse to do what I required him to. Yet he strove to free himself, and was probably glad when I let go. Joan was shorter, being barely six inches long. Every little while these two would lift their heads suddenly up to the surface of the water, or above it, as if to snap up something, and immediately a bubble or two of air would appear on the surface.

Thirteen days after Darby and Joan first entered my dish, a girl came to see me. She is one of the few persons of my acquaintance who are interested in "bugs," and she had come in to expatiate on the beauty of a dragon-fly larva of hers that had just taken off its skin, and presented a beautiful, velvety appearance. From dragon-flies the talk whisked to the small tortricid moths that, as caterpillars, curl the leaves of rose-bushes, and I invited her into the back yard to see if any more of my tortricid pupæ had opened.

While there I was reminded of my Water-lizards, and proceeded with some enthusiasm to untie the cloth in order to show Darby and Joan to the visitor's admiring gaze. Off came the cloth. Nothing was to be seen of Darby and Joan, but that was not strange, since they were given to hiding themselves under the grass. So I pulled it aside and took out some of the stones. The polliwogs wriggled blissfully, but no Darby and Joan appeared.

"I guess they're in the cloth," I said, with an uneasy foreboding of evil.

I shook the rag.

Alas, alas!

The truth was apparent. Darby and Joan had run away.

How they had managed it was a mystery. I had tied the cloth on as usual. Perhaps it was a little loose, and it may be that by great struggles those creatures, almost as pliable as slugs, slipped through and fled.

The girl went away almost immediately, and I rushed back to the dish. Away went the box of bottles and tins of insects, up came the heavy boards; into every crack and cranny among the sweet alyssum did we look, but a dried Darby or a parched Joan appeared not.

I am quite certain that Joan never planned that escape. She need not come back to apologize. I exonerate her entirely. She was of too meek a nature to propose such a thing, or else she was a tremendous hypocrite. I know well enough that it was Darby the daring that found the way out and persuaded her to follow, but, inasmuch as there was no water in the open world to which he invited her, I very much fear that Joan's fidelity was rewarded by a dry death.

Still, I am glad to have become acquainted, though so briefly, with Water-lizards. It is something to have even seen an animal, as poor Paolo

Uccello might testify from his own mistake. Alas, poor man! Being at work decorating the arch of the Peruzzi, he placed in the rectangular sections in the corners one of the four elements accompanied by some appropriate animal; to the earth a mole, to the water a fish, to the fire a salamander, and, since the old notion was that a chameleon lived on air, he was to paint one of those creatures in the right place for that element.

But Paolo had never seen a chameleon, and, being deceived by the similarity of the names, what did he do but paint a camel with wide-open mouth swallowing the air. "And herein," says Vasari, "was his simplicity certainly very great; taking the mere resemblance of the camel's name as a sufficient representation of, or allusion to, an animal which is like a little dry lizard, while the camel is a great ungainly beast."

CHAPTER VIII.

MINOR MUD AND WATER FOLK.

"By being seldom seen, I could not stir
But like a comet I was wondered at."
King Henry IV.

A WHILE ago I heard of a heroic act of self-sacrifice. A neighbor of mine, a good-hearted youth, is an enthusiast about bugs. He is also a member of that institution of the nineteenth century, the Salvation Army.

One evening, when he with some other members was holding a street-meeting, and was just in the act of kneeling for prayer, the eyes of this person fell on a bug, or beetle. Perhaps it was an emissary of Satan to tempt my neighbor, but at all events it was a new bug to him. He could not have helped seeing the insect in the light on the street, but it was an awkward moment for such a discovery. Should he seize that bug, or remain in a reverent attitude?

Devoutness conquered, and the bug escaped. Perhaps a scorner of bugs might think this a trivial act of self-denial, but a bug-hunter could have given no better proof of religious earnestness.

"Do you really think they're alive?" said the same person to me, when I met him on the hill one day and offered him my jar for inspection.

I had been making a call on the Mud Folk. There are a good many of them beside this brook. They are not wont to receive my calls with enthusiasm, neither do they press me to come again. But I dig into their habitations without ceremony. The question of my neighbor referred to a long hair-worm, *Gordius*, that I had taken from his mansion in the mud and was conveying home.

On my affirming my certain belief in the life of the creatures, the lad said, "But folks say they're horse-hairs. I used to see lots of them. They're not all black; some are red. I used to think it was because they were different colored hairs. Did you ever try putting horse-hairs in water and letting them come alive? I used to do it when I was a little fellow, but I never made much of a success at it."

I should think not! Poor *Gordius aquaticus*. Will human beings *never* become tired of repeating that old story of the horse-hair?

In an evil hour I resolved to experiment on that *Gordius*. I was moved thereto by reading a statement in some book that certain hair-worms can be dried into brittle threads and yet will become active on being moistened.

So I put my friend into a dry bottle and left him for about a day. He was then a miserable little ball with no apparent life, but I gave him some water and he was soon stretching himself in it. He was unmistakably alive.

Rejoiced at my success I allowed *Gordius* only a few minutes' happiness. I resolved that I would try him with a longer drying spell. Two days should be the limit this time; and into the driest of tin cans went poor *Gordius*.

I think it was a few hours over the allotted time when I remembered my captive. He was wofully dry, but I had hopes of him and put him into his bottle of water.

Gordius aquaticus.

But it had been too much of an experiment for poor *Gordius*. His body became plump and round, but there was no life in him. Even when he had stayed in the water over night and all the next day, he was motionless and allowed me to measure him, a thing that he refused to do when alive. He was about thirteen inches long. I had thought that when alive he was considerably over a foot in length, but perhaps his drying shrunk him.

There are numbers of big flies sitting on that margin where the sun shines so warmly. Are they not also Mud Folk? If St. Macarius lived in these days, he might come here for flies, though they have never bitten me. Perhaps I have never given them the opportunity that he gave such creatures in those marshes of Scete which contained multitudes of huge flies "whose stings pierce even wild boars." For the legend goes that feeling a gnat bite him one day, he, like any ordinary mortal, killed the creature. But, immediately, so they tell us, his saintly mind was overcome with remorse at having lost so good an opportunity of mortifying the flesh. With zealous haste he rushed from his cell to the marshes, and there, amid the flies, abode half a year, and when at last he returned to his friends he was so disfigured a man that he was to be known by his voice only.

In view of such self-mortification as this, it is reassuring to ordinary mortals to remember the other legend of St. Bernard, who is said to have become one day so annoyed by a blue-bottle fly that buzzed about his ears that he said, peevishly, "Be thou excommunicated," and lo! the flies of the whole district dropped dead.

These big flies are not extremely interesting objects. Nothing about a fly is, to my mind. He is associated with too many unpleasant recollections for me to rejoice in his children or in any of his relations. He is an insect known to all people.

On a fence that walks up a hill I once found

in August a number of big Horse-flies sitting in the sun. I clapped my tin over one of them, but so anxious was he to get away that I bruised one of his wings before I could draw my handkerchief tightly between him and the fence. He was big and black, and his thorax was crested with a stout-looking brownish shield that covered him like a breast-plate put on the wrong side. He buzzed like a bumble-bee when he flew, but when examined he was seen to differ in having no hairy body but a deep-black one. It is only the female Horse-flies that bite, it is said, the males, like those of the Mosquito, living on the juices of flowers. The insect creation serves sometimes to emphasize the opinion of the ages as to woman's temper. I passed a number of cows back on the meadow, but the grass prevented any of them from being thin enough to represent that ungallant fiction of the French mind, the "*Chichi-Vache*," or "sorry cow,"—a monster that was said to be exceedingly thin. Its diet consisted of good women only, and the "Chichi-Vache" was all skin and bone, because its food was so extremely scarce, such females being very rare.

When you raise the mosquito-bar that covers the top of the fly-larvæ tin, little black specks fly out. These on being looked at closely resolve themselves into minute flies that are hardly bigger than the small black ants that visit the tin. Small beetles, and beetle-pupæ of the *Staphy-*

linidæ, are also in such mud beside the brook, and earth-worms abound. The little black beetles are determined not to be captured, however, and even if you put half a dozen in your tin and pat down some earth over the mud, up come the beetles through the earth and straight to the top of the tin they go.

Perhaps another time when you raise the mosquito-bar of your tin another fly, about the size of a small house-fly, stands waiting for release, though whether he came from the mud or slipped under the mosquito-bar from outside is a question.

But in all my digging in the mud I have never found any swallows, and have never verified the statement that truthful Mr. Harrington made to Mr. Pepys in 1663 that December day in the coffee-house, " Swallows are often brought up in their nets out of the mudd from under water, hanging together to some twigg or other, dead in ropes, and brought to the fire will come to life."

Yet I venture to say that many people could not tell me the names of the real Mud Folk, and the guesses might be as false as the swallow-story that was told of the " country above Quinsborough."

Innocent Mr. Pepys! He seems to have believed all that was told to him. I am afraid he did not go poking around in the mud enough when he was a boy. I should think that he was the sort of man who could be deceived by the imitation-insects said to be manufactured by deceptive bug-dealers for the cheating of the unlearned purchaser.

Among the more minute inhabitants of this brook are the entomostracans, known as Waterfleas, *Cypris*, *Cyclops*, etc. In a small puddle containing two or three teacupfuls of water you will see swarms, hundreds of *Cyprides* in their two-pieced horny shells. *Cypris* is extremely lively in the water, but take him out and put him on a glass slide in a drop of water and he usually keeps those finely-fringed antennæ and feet inside his shell where the would-be see-er cannot look at them. Occasionally, however, he protrudes them, and whirls around at a great rate, demanding to be put back into the water where it is deep enough for him to swim.

Cypris unifasciata.

Cyclops communis.

There is nothing that will make the wrinkles come more quickly than daily squinting at such creatures as these. They are so small, and yet one wants to see their swimming feats, and continually puts one's eyes on the strain to observe them. And when you mix them with a number of the genus *Cyclops* there is perpetual commotion. I once had a jelly-glass of water full of such creatures, and in the time I kept them I nearly obtained one or two wrinkles in the middle of my forehead. Any one who wishes to cultivate wrinkles will please take notice that squinting at Water-fleas is the surest and quickest way I have yet discovered. But it is interesting work to watch the entomostracans through the glass. The lady *Cyclops* will shoot by bearing two egg-masses nicely balanced on either side, and *Cypris* with all his brethren will skim around over the floor of the jelly-glass looking at a glance like a lot of little spiders in for a swim. Multitudes of the fossil shells of the *Cyprides* are said to be found in the Wealden rocks of England, in the limestone of the carboniferous series, etc. Such little creatures make more marks on the earth sometimes than men do, — a fact that is sufficiently humbling to human pride.

Another quite lively and brilliant little creature that I have found here in March is a "water-mite," one of the *Hydrachnidæ*. Bright red is this mite, so much so that it looks almost like an animated speck of blood as it goes through the water. Its color forms a strange contrast to the usual tints of water-creatures, which are commonly dark, or yellowish white.

This red mite has eight legs, and in shape is much like a small spider. To look at the mite, one would hardly think that it belonged in the water, but it seems much at home in that element, and I fear that when in the larval state the creature is too much at home for the comfort of

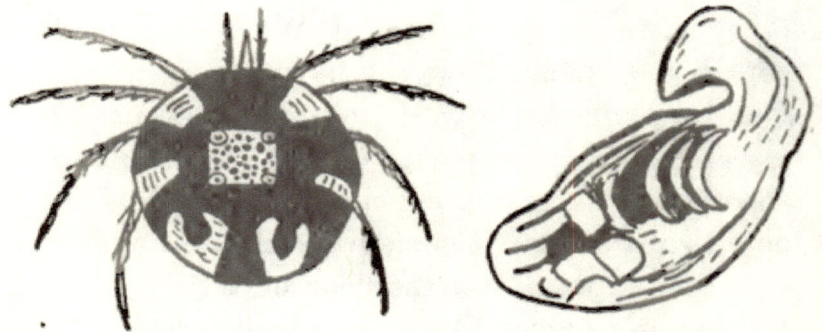

Water-mite Adult.   Water-mite Larva.
Hydrachna geographica, — magnified.

the other water creatures from which it takes its food. For some of the larvæ of the *Hydrachnidæ* are said to be parasitic on the gills of mussels, and other larvæ hang on the Water-skater family, — the *Hydrometridæ*. Ranatra and the Water-scorpions are also said to be bearers of the nymphs of these pests. Some of the Water-tiger beetles, *Dytiscidæ*, are devourers of the adult mites, however, so justice is sometimes meted out to these evil-doers. Between the leeches and the water-mites, one might think that sometimes the life of the bravest bug or beetle in the brook would become intolerable.

You sit down on a bank and think yourself alone. But there are many eyes on you. You hear a rustle, and behold a little white frog is hopping away from your feet. Aphides alight on your shoulder, and now would that a person might have as acute hearing as the mythic Heimdall had, he who lived in the fort at the end of the rainbow, and had so fine ears that he could hear "the wool growing on the backs of sheep and the grass springing in the meadows." For, if we had such ears we might hear many a bug and beetle and spider in this supposed-uninhabited nook saying to each other, "Who is the giant who has come here and why has he come? This part of the world belongs to us. We are the Brookside Folk."

And you reflect that the spiders and beetles and bugs really have more to substantiate their claims than had those "Three Tailors" whom Carlyle mentions as addressing Parliament and the Universe, sublimely styling themselves, "We, the People of England."

Ants walk beside our pool on the mud near the margin. What may be their errand there I do not know, for there are no live-oaks or willows in that particular spot to attract them. Where willows grow you may find the backs of the leaves brown with aphides that the ants are visiting. The Portuguese of Brazil call the ant the King of that country, and the peasants of Cornwall say that ants are " Muryans," or small people of the fairy tribe, " in a state of decay from off the earth."

Whatever they are, they boldly intrude into a bug-keeper's tins. I have, in feeding my caterpillars with leaves, carelessly put in some that had aphides living on them, and on looking in once I saw two ants, each carrying an aphis in his mouth. Evidently the ants were going to bear the aphides to some spot that in ant-judgment was better adapted for their cows than my tins were. I doubt, however, whether the ants could squeeze through the mosquito-bar with such burdens.

The black polliwogs come very near being Mud Folks in April. One will find the water black with a quivering company of them next to shore, or directly on it where the water just covers their backs from the sun's rays. I am afraid some of these black polliwogs miscalculate sometimes, or are thrust by some cruel fate a fraction of an inch out of their element, for one can sometimes see them above water-line, miserably waiting till the sun shall dry them up or till some kind hand shall put them into the water again. If the latter, their passiveness vanishes, and they go off through the water, shaking themselves with happiness. I wonder if, when those unlucky Lycian shepherds who mocked Latona received their punishment of being turned into frogs, it was decreed that they should pass through the polliwog state? That would have been too cruel, to condemn mortals, who had been used to having feet and using them to escape from dangers, to a footless condition wherein they would be comparatively helpless. I am sure a tadpole must feel proud of his first pair of feet, albeit they are his hind ones.

Strings of eggs are toad's; round eggs are frog's, say the books. I remember finding here, one March, a long string of double black dots enclosed in yellow jelly, and taking it home I had the pleasure of seeing the dots lengthen till the polliwogs' tails stood out of the mass in all directions. And finally the whole polliwogs would slip out and sink down through the water in a half-alive way, as though they really did not know whether they were glad to be in the world or not. They gather courage after a time, however, and conclude that it is a fine world, after all, and it is a blissful thing to waddle in it.

I found here two boys inspecting a can of theirs one afternoon. They were seated on a bank, and their can contained a collection of little frogs, etc.

"What do you do with them?" I queried.

"Have fights and see which side whips," explained one youngster.

And forthwith, aided by his companion, he made clear to my comprehension the joyful manner in which the fight was to be conducted.

"The side that whips we're going to give that pond," said one, pointing to quite a pool.

"And what becomes of the side that gets whipped?" I asked.

"You can't do much with them, but throw them away. They're mostly dead, anyhow," was the answer.

And then they explained to me that different kinds of "bugs" were on opposite sides and the fight took place in a tub of clear water.

And the boys solemnly assured me that "this kind of bug" (pointing out a water-shrimp that was walking on the black mud) attacked the polliwogs.

"Guess they suck their blood," said one boy.

He stated for a fact what I have never noticed a shrimp doing, but what I should not be at all surprised to see such a crustacean do, that one of them will catch hold of the polliwog by the throat and kill it.

"They don't do it in the pond, but when they're shut up together they do," said one boy.

"It's fun to see them fight," said another of the youngsters.

And having obtained as many creatures as they wanted the boys ran off.

But those creatures on the "side that whips" must think they have won a glorious prize, the pool that is theirs by good right anyway.

Here may one find in March the larvæ of Mayflies with branchiæ on their sides and slender bristles behind. These creatures seem to have an unaccountable tendency toward death. They appear very lively as they skip along through the water after their pretty fashion, but take them home and try to keep them until transformation and you will be likely to make a failure with most of them. Heartless things they are, too, capable of hustling about the dead bodies of their brethren in a most unfeeling manner. It is well that death does not frighten them, else their lives in a pool would be full of terror. If I were one of the water-creatures I should constantly be expecting my own demise, though I should keep my eyes open, not shut, as did Cosmo de Medici, of whom it is related that, a short time before his death, his wife asking why he kept his eyes shut, he replied, "to get them in the way of it." A water-insect having no eye-lids could hardly follow the Italian's example.

Multitudes of mosquitoes must go forth from some of these pools in the course of the summer. It is strange that there is hardly a bit of land or water that is not claimed by some of the lower creatures. We do not know it, perhaps, but we learn it after a while. One may collect insects in one's own neighborhood for a long time, and yet may be astonished at finding new ones in the next hollow. So, after a time, one comes to believe Charles Kingsley's saying, "He is a thoroughly good naturalist who knows his own parish thoroughly."

But one does not need to be much of an entomologist to be acquainted with the mosquito, although acquaintance with the eggs of that insect is a different matter.

The mosquito was the first boat-builder. Of that I am certain.

> "Vessels large may venture more,
> But little boats should keep near shore,"

says wise B. Franklin. Mrs. Mosquito has probably heard of that bit of wisdom, for she sometimes chooses such very small puddles in which to sail her boats that a careless cow, stepping in, tramples the puddle out of existence, much to the detriment of mosquito children, no doubt.

You will take Mrs. Mosquito's boat for a bit of soot, if you are not careful. But pick it up and you will see the tiny eggs massed together so regularly, the whole narrowing toward either end according to boat-shape. You cannot make such a boat sink. Put it into a bottle and pour in water. The mosquito-boat will always come up on top, if the eggs have not hatched. Mosquitoes vary in the length of time they require ere they can come forth to bid defiance to mankind. Some that I had took sixty-eight days or more, I believe, but those mosquitoes must have been quite a dilatory set, the usual time being about four weeks after hatching. Mine were a fall brood, however, hatching from the egg October 3, as animated whitish commas with black heads.

Mosquito infants are really bright-appearing persons. A good many people are not aware that there are two forms of these creatures before they have wings. A man with whom I once conversed, while knowing mosquito "boats" and mosquito "wrigglers," was yet ignorant that there is a middle pupa stage of "tumblers," as the club-shaped creatures are called. The tumblers are said not to eat anything, so one can hardly wonder at the hunger of the winged creatures that come from them. Well do I remember spending a night among the swamps of the Sacramento River, when a child, and seeing the enormous swarms of mosquitoes above the trees, humming like a lot of bees. Wise writers say that the lady mosquitoes alone bite, and that the gentlemen content themselves with dwelling in swamps and woods. But four trustworthy people and two veracious horses could testify that there were a large delegation of lady mosquitoes in the Sacramento swamps that night, and that fire and smoke did not prevail against them.

The Flamen Dialis, priest of Jupiter, was anciently not allowed to touch a trailing vine. The poison-oak along this creek seems to be determined to enforce the same rule on everybody as far as these blackberry-vines are concerned, for the two plants here grow together, and the would-be inspector of blackberry leaves has to be very careful lest he should lay hold of the wrong plant.

But if, on the last day of April or the first or second of May, one has patience to edge one's way along the bank, holding to the fence with one hand and stretching out the other, grabbing with judicious care, one may reach berry-leaves done up in a suspicious style. Pick them and open the leaves so bound together or folded, and you will find within each bundle a small green worm with a darker head, one of the *Tortricidæ*, or *Pyralidæ*, for I believe authors differ somewhat in regard to the two families. At all events such worms are destined to become pretty little moths some day.

But, if you find the right leaf, on its back you may have the delight of seeing a family of little bugs just hatched and sitting together on top of the eggs they have come out of. I have found the same kind of eggs on the back of a honeysuckle leaf in my yard. There are about fourteen eggs in a group, and they are very pretty, barrel-shaped, of a pure white, but marked with red before the inhabitants come out. Hardly one sixteenth of an inch high are the little barrels, looking a little like the pictures of the egg-barrels of the Harlequin Cabbage-bug, *Murgantia histrionica*, or the Calico-back, but lacking the black hoops and bung-holes of the eggs of that

enemy of farmers. But there are lids to these blackberry-leaf barrels, or at least there is a rim around each top edged with little hairs, and when the young bug has cut its way out with a neatness and accuracy that could not be surpassed, behold the "cutest" white
Eggs on back of blackberry leaf, — enlarged.
lid in the world stands just a crack open on top of the empty, almost transparently-white "barrel." You can take a pin and open the lid if you do not push too hard. If you are not careful, though, it will come entirely off. There seems to be a sort of dark projection on the middle of the open side of the lid. Perhaps it is something to keep the lid part-way open when the young bug is crawling out. Who knows?

The eggs are placed in quite regular rows. I have found the fourteen in rows of the following numbers: three, four, four, three. In another case the fourteen were in this order: four, four, four with the second one raised above the others, two.

The young bugs that come from such eggs are of necessity small. They are six-legged, with broad ending antennæ, and the front half of each body is reddish brown, while the hinder half is lighter with four brown marks, and a row of brownish dots forming an edge to the body.

Little bug a day old, — enlarged.

These less than pin-head-size larval bugs have

brave spirits; for a day old fellow climbed clear to the top of the glass fruit jar in which I kept them, and if I had not had paper tied over the top he would have gone forth to seek his fortune.

CHAPTER IX.

CADDIS-WORMS.

> "Sometimes he angers me
> With telling me of the moldwarp and the ant,"
> "And such a deal of skimble-skamble stuff
> As puts me from my faith."
> King Henry IV.

In the other brook, on stones in the water, yet near the margin, I have found small Caddis-worms in their tubes. Among those that I once brought home was a little fellow barely three eighths of an inch long. He had made his tiny case of grains of sand and minute stones very neatly put together.

For a time I gazed upon him, and then commenced the struggle of Sandy's life. Very carefully I began to take his covering to pieces, intending to see him make a new one. But so tenacious were the materials that I only succeeded in making a hole in the hinder portion, leaving a little of his body exposed. I was too much afraid of hurting him to go on.

One of my cases with Caddis-fly larva in it.

But I had done enough. Sandy knew the dangers of having holes in one's clothes. He would attend to that rent immediately.

I handed him a little stone; he clutched it and proceeded to undress, regardless of spectators. His covering was a very tight one, however, and it was very hard work. I think he would probably have succeeded in his efforts without my aid, but I "lent a hand," as Mr. Hale suggests. The dress was so tight around the neck that Sandy looked as if he were choking. But at the end of almost half an hour, Sandy caught hold of another caddis, I caught hold of Sandy, and he drew himself out.

There he was, a poor, naked, yellow worm in the waste of mighty waters contained in a small plate.

Now Sandy was in business. I had thought that he would be so exhausted he would lie down and rest after taking off that garment. But no. He was immediately at work. How did he know but a fish might be coming to swallow him before he could get covered again? There was no time to be lost, in his opinion.

Poor Sandy! How he did toil and tug over the little stones. His black, ant-like head and yellow body with its two terminal hooks went squirming around under and over the stones. I did not understand his movements at the time. I thought he was debating whether he should make his new covering out of stones or of sand. I had destroyed his former garment so that he was obliged to make an entirely new one.

But I saw afterwards that what he had been doing was to anchor himself for his future work; since subsequently, when he was making his new coat, he was so attached to the larger stones that I could pull him around by moving any of the half-dozen around him. They seemed to be all connected by some thread-like substance, and I thought it a very sensible idea of Sandy's, for it would have been awkward work dress-making, I suppose, without a certainty that he could keep in a fixed place.

Being anchored, Sandy worked. It lacked a few minutes of seven hours after this when Sandy slipped down his body the first section of his new dress. He had seemingly made this thin section round about his body near his head, and then he caught hold of a little stone that I had pushed near him, and with mighty squirmings slipped the section down toward the other end of his body. This was probably the "trying-on" part of the dressmaking. The garment was to be a "glove-fit," without wrinkles, hence Sandy's feats in wriggling.

In making the section he seemed to pick up the sand-grains and manipulate them; but Sandy need not have been afraid of my stealing his trade, for though the operation went on under my eyes, I could not tell at all how it was done. The first section covered nearly half his body, exclusive of his head. He manipulated a while longer, and drew the section down further still. I had been thinking that he had not drawn it quite far enough, for I was almost sure I saw one of the hooks at the end of his body peeping out. That showed afterward, too, for the section went up near Sandy's head again, and left his rear half bare. But Sandy knew what he was about. He had made a dress before. He caught hold of the stone and shoved the dress down his back again.

Then I went to bed and left Sandy to his job. It was "quite a chore," as a neighbor of mine says of other things.

When I arose the next morning, Sandy apparently had not made much progress in the length of his garment. He lay on his back some of the time that day, and rested from his toil, I thought, but the next morning revealed a sad state of affairs. Poor Sandy had spun his little life away. I brought him a bit of grass as refreshment, but it was too late. He was quite stiff and dead when I tried to turn him over. His dress covered about two thirds of his body, reaching well toward his head on his back, but I never had the opportunity of seeing what was his intention about unmooring himself from the larger stones, when his dressmaking was done. Perhaps he had no such intention.

Sandy's unfinished "chore."

My caddis cases varied in length from seven sixteenths to three quarters of an inch, the latter length being represented by but one, and due to a projecting stone. One case that was a quarter of an inch wide in the broadest part was weighted by seven stones, while the longest case was a mass of them. These cases were closed and contained pupæ. When the little sand-dressed larvæ took their walks up the sides of the bottle and on the leaves, the creatures reminded me of small hermit-crabs peeping out of their shells. The Caddis-worms seemed to have much the same lively, impertinent natures as those crabs, too.

Once in April in the other brook, when a great

Papilio Turnus.

yellow *Papilio Turnus* butterfly was flying up and around the water-side, there was a bigger

Caddis-worm than Sandy had been. This larger Caddis was suddenly torn from the home of his infancy. No more should he wander through a pool dragging sticks after him. His home, when it was not a plate, was to be a bottle.

This Caddis-worm was about one inch and three eighths long while in his case, because he had hitched two small logs to himself and insisted upon dragging them about with him. He was a lively fellow. Perhaps dragging weights is a good thing for a Caddis-worm's constitution. Anyway he appeared so healthy that I thought he could endure making another covering for himself, and not die in the operation, either.

The bigger Caddis-worm's House with "logs."

So I took his garment off from him and commanded him to go to work. I pounded up some red brick and some white and blue buttons and gave him the pieces, expecting him to make a patriotic-looking garment.

It was about a quarter of two P. M. when this worm began dress-making, and his house was finished by the next morning. It was evidently hard work and he was inclined to keep pretty still the next day, instead of trotting around the dish as he had done before a new dress was demanded. In making this dress he did indeed attach to himself a piece of red brick and one each of white and blue buttons, but his immediate covering was mostly of sand, and on his bosom he had fastened quite a sprig of green grass. I had laid that grass before him, thinking that perhaps he might like a bite during his exhausting labors. But E Pluribus Unum had Hibernian tendencies, evidently, and believed in the " wearin' of the green." He reversed his position, however, so that he afterward wore the grass at the other end of his garment.

I had thought that he was going to prosper finely after his dress-making, but two or three days afterward I came to his bottle and found him dead. The cause was overwork, undoubtedly, and no matter how vigorous a Caddis-worm may seem, I do not think that I shall ever impose such labor on one again. Dress-making is evidently a great exhaustion of the vital forces of a Caddis-worm, and I felt guilty as I reflected that I had murdered two of the inoffensive beings.

In some places in that brook there are so many short, pebble-covered caddis cases on the stones that the gazer is irresistibly reminded of the little barnacles seen in such numbers on the rocks by the sea.

I brought home one Caddis-worm that was certainly an ingenious fellow. I marvelled at his style of architecture. It was somewhat the same that children use in building " houses " of clothespins. The little sticks stuck out in all directions, and the gifted inhabitant of the structure looked very queer, hugging the stalks of grass on which he was wont to hang in the bottle. He must have had to hold on tightly, with all those sticks weighing him down. Still, he sometimes hung perpendicularly suspended by his legs. He could manage his house pretty well, and often made it stand out in different directions to suit him as he climbed. I should not have had the heart to take to pieces his house. It represented too much labor.

A triumph of architecture.

The little sand-covered Caddis-worms were quite given to hanging from the grass. I remember seeing one " shin down " a slender stem, until losing hold, either intentionally or not, he floated swiftly and gently down to the bottom. I do not know how any naturalist could see one of these little beings weaving so deftly its sandy dress, and not recognize the existence of the Infinite Kindliness that watches over even the meanest creatures and provides them with means of protecting themselves against the little bruises that must be so great to them. I do not know whether Job ever saw a Caddis-worm, but I think that the man of the land of Uz never gave his friends truer or more beautiful advice than when he said to them:

"But ask now the beasts, and they shall teach thee; and the fowls of the air, and they shall tell thee:

"Or speak to the earth, and it shall teach thee: and the fishes of the sea shall declare unto thee.

"Who knoweth not in all these that the hand of the Lord hath wrought this?"

Open some of the barnacle-like, stone-covered little caddis cases that stud the stones and you will find the pupæ. One that I examined was about a quarter of an inch long and showed very plainly the black eyes of the coming Caddis-fly.

These flies should remain about two weeks in a pupa state, I believe. Perhaps the water in a glass bottle may not be fresh enough for the pupæ. At least I have never been able to keep my caddis pupæ in them until transformation. The pupæ seem to die, and a slight mould, such as one observes gather on all dead insects under water, forms on the caddis cases.

CHAPTER X.

MY CORYDALUS.

"Mislike me not."
Merchant of Venice.

CORYDAL*IS* or Coryda*lus*? Authorities differ. Let us "follow our leader," Packard, and say Coryda*lus*, despite the array of Tenney, Duncan, and Wood in favor of "is."

Never mind how Corydalus spells his name. He is an evil beast.

I well remember the day when my first and only larva of Corydalus was found in the further creek. I have cause to remember it, for did I not find some caterpillars dwelling in closed nettle-leaves, and did I not sting myself most unmercifully in taking possession of my findings and pulling up part of the nettle to carry home?

Did you ever prick your hands beautifully with nettles, and then wash in a brook underneath willows? Ah, that is a sensation! One hardly wonders that in olden times people used nettles for striking paralyzed limbs, hoping to bring feeling back into them. What callous hands the old Germans must have had in those days when they made the cloth called "Nessel-tuch" from nettles.

What is the reason insects care so much for prickly things? To be sure, it is very well for them to do so, as the prickly plants are often those that are called " weeds " by men, and that are considered obnoxious. Surely it is better that caterpillars should feed on them than on cultivated crops. Still, one cannot but reflect that there are many weeds that do not have spines, and one wishes that caterpillars would be satisfied with them. But the bug-hunter must not capture caterpillars unless he expects to get them their usual food.

Yet, when one possesses a caterpillar that desires to make a repast on spiny thistles, one is apt to have a slight bond of sympathy with that historic gentleman mentioned by Mother Goose: —

> "Simple Simon went to look
> If plums grew on a thistle;
> He pricked his fingers very much,
> Which made poor Simon whistle."

When a person is bringing up a number of caterpillars that feed on nettles, with a family of bugs that demand blackberry leaves, that person's fingers are likely to pass into a state of prick almost unendurable. I know whereof I speak, for have I not gone with blistered finger-ends to feed my unsatiated caterpillars more nettles? It is cheering, when in such a plight, to remember that old writers used to recommend stinging with nettles " to let out melancholy."

From this, the observant reader will perceive that I have never yet learned to grasp nettles firmly enough, a lesson inculcated in the old rhyme written by Aaron Hill in 1750 on that window in Scotland.

> "Tender-handed stroke a Nettle,
>   And it stings you for your pains.
> Grasp it like a man of mettle,
>   And it soft as silk remains.
> 'T is the same with common natures,
>   Use 'em kindly, they rebel;
> But be rough as nutmeg-graters,
>   And the rogues obey you well."

That day I overheard a cry from my companion and rushed to the rescue. There in the shallow stream, doing his very best to get away, was a full-grown larva of Corydalus. He was popped into the bottle without ceremony and brought home. In appearance Corydalus was no beauty. I presume he would have frightened some people out of their wits. But I was overjoyed to see him, for he was the only one of his kind I had ever beheld outside of a book.

He was about two inches long, had a black head with nippers, six legs in front, and eight respiratory filaments standing out from either side of his body and making themselves so conspicuous that an impertinent little Caddis-worm that wore a dress of sand caught hold of one of them. I presume Caddis pinched, for Corydalus turned around on him with the same quick motion that you see in a cat or dog suddenly bitten by a flea. Caddis held on impudently till Corydalus had several times turned on him in wrath. Then I interfered, and Corydalus took Caddis in his nippers, but the pinching rascal retreated into his case and was safe from well-deserved vengeance.

Corydalus was "pudgicky." He took it as an insult if I managed to touch him, and he climbed recklessly around, disdaining the gift I made him of an earth-worm. Corydalus despised me; nay more, he hated me. He would have none of my favors. Was I not his jailer?

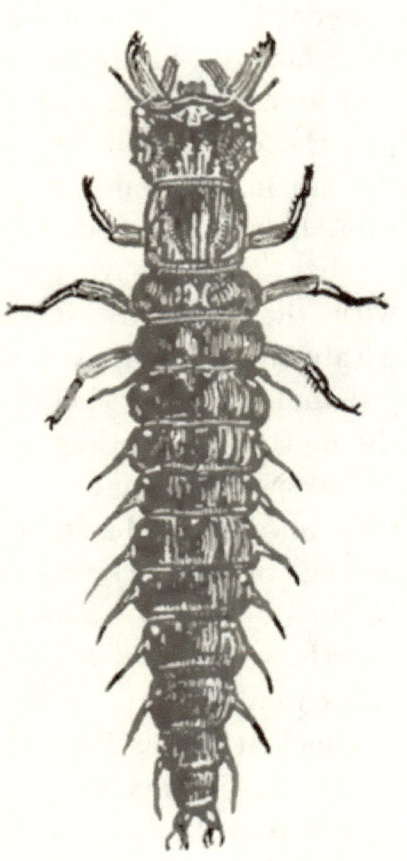

Larva of Horned Corydalus.

I gave him a little mortar for a pond and set it inside a flower-pot of soft earth, so that he might climb out of the water if he so desired, for I thought from his size he must be near the time of his transformation.

The pot was too full of earth, and Corydalus in his customary defiant way plunged headlong off into the air and was hurled down the precipice to the table. Such adventures merely agreed with Corydalus' temperament. I fancy that he passed the whole of his life in that sort of defiant humor that sometimes makes a foolish human being reject the comfort of his fireside and rush out into the storm, rejoicing in whirlwinds and pouring rain, merely because they furnish something to combat, and the very force exerted in battling with the elements serves to quiet the tempest within the one who uses the force.

I took out a portion of the earth and lowered the mortar, replanting the grass around it. Corydalus speedily came out, and, after some hesitation and going back to the water, established himself in the darkness and dirt under his lake. He was evidently no longer to inhabit the water, and the poor earth-worm might rejoice, for Corydalus would never eat him.

I looked in on Corydalus several times. The last time he was so indignant that he plunged deeper into the earth and I let it fall in so that he was buried from sight in the flower-pot. I carefully kept the hole closed in the bottom of the pot, however, for I did not want him to escape that way. He burrowed under the earth, as Rabbi Simeon declared the just do. That truthful rabbi asserted that for those righteous persons who were buried outside the land of Canaan there should be caverns made beneath the earth, by which the just might work their way till they came within the sacred limits of the land of Israel.

Truly a wonderful amount of burrowing goes on underneath our feet, but the burrowers are far from being those persons of whom the rabbi dreamed. Neither should I apply the epithet "just" to Corydalus. His appearance was against him.

The rabbins say that the names of the angels were first learned by the Jews during their captivity. But Corydalus, during his, had no desire to learn the name of even the person who attended him. Perhaps he did not look upon me as an angel. Very likely he did not. And he would not have flattered me by insinuating that I was one, if he did not think so. He was not that kind of a person. Still I was happy to have found him. It is not on every journey to that far-away brook that one can find such a treasure. Do not I remember walking there one morning and being so frightened by a dog and a cow that I fled homeward bearing with me but a few miserable caterpillars and a larvel Frog-hopper? How much happier was the journey during which Corydalus was found.

In the other brook one May morning I found floating dead an enormous Mole-cricket, one of the creatures called by the French "Courtilières," from the old French word *courtille*, "garden." I once kept a Mole-cricket in a bottle for quite a time. He used to ascend to the top of his hole occasionally in the twilight, but he was a timid young fellow, and never became acquainted with me.

It is hard to wait for such a creature as Corydalus. He disappears in the earth, it falls over his much-branched body, and you see him no more. Days go by, fifteen, sixteen, twenty of them. You grow nervous. You wonder what he is doing in the lower regions of that flower-pot. You feel possessed to dig down and see him. You know exactly how the little boy feels when he has planted his first seed and longs to dig it up and see if it has begun to grow. You hunt up Corydalus in books, and you find Hagen saying: "The reason that the larva of Corydalus has both branchiæ and spiracles is, that it lives, like *Sialis*, some weeks out of the water before its transformation."

You heave a sigh and shut the book. The gloomy thought oppresses you that perhaps Corydalus has not transformed into anything. Perhaps he looks just as he did when he last saw the light. It is very discouraging, but you remember what damage was done to the Water-tiger larva by digging down to see him, and you bide your time, or Corydalus' time, rather. You console yourself by reflecting that at least Corydalus did not require you to dig his hole for him. A boxful of these creatures would be much like that monastery of the Catacombs on the Dnieper where some of the ascetics are said to have

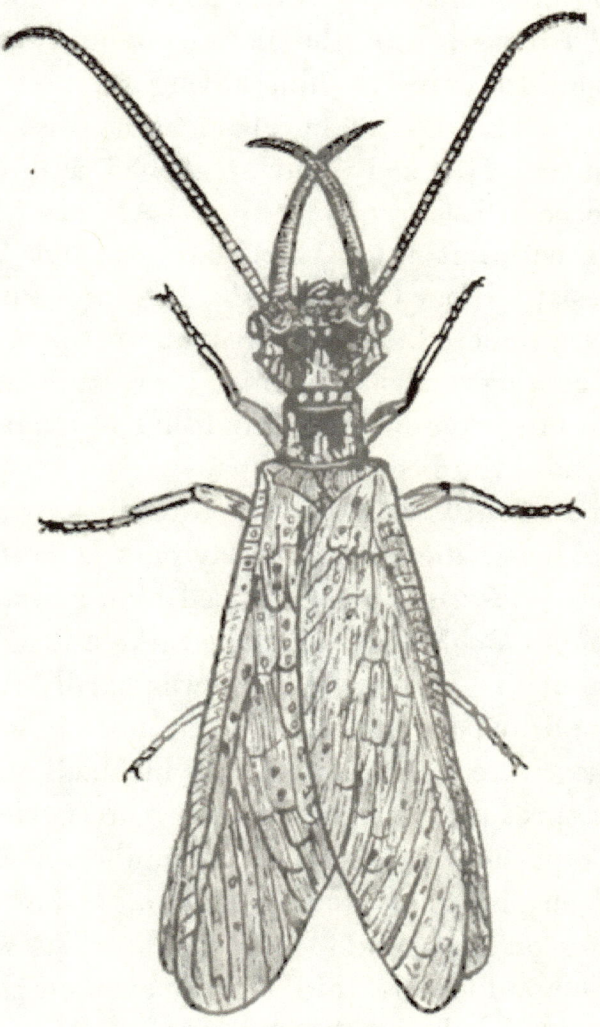

Horned Corydalus.

bricked themselves up alive in the cells which were to become their sepulchres. But Corydalus is wiser than the saints. He intends to come out some day while he is yet alive.

On the twenty-second day after Corydalus was found I looked into the flower-pot and was astonished to perceive him sitting on top of the earth. But, alas! how changed he was in appearance. His body had shortened and he was altogether discouraged-looking. All his fire and wrath had left him. I poked him, but he did not resist. Poor Corydalus! His days were evidently numbered. There was no prospect of his ever coming out as a winged creature with a couple of savage-looking mandibles in front. He could not even become a pupa.

He evidently came up on top of the earth to die. He did die, and as he lay on the bottom of the jelly-glass in which I placed him, he had contracted so much that he looked like a dead larva of one of the *Dytiscidæ*. He was hardly an inch long and his side-appendages did not show much for they were drawn up against his body, and in this respect the resemblance was more perfect.

I kept him in the glass a number of days to see if any internal parasite that might have been preying on him would come forth. The wicked Ichneumon-flies are said to prey even on Caddis-worms, and I do not know why Corydalus might not have been attacked some time when he chose to walk out of the water. But no parasite appeared, and I was somewhat at a loss to account for his decease.

Still, it must require much strength to pass through the transformation-period, and perhaps Corydalus was not so strong and healthy as he appeared when I brought him from the brook.

I think that if Corydalus could have chosen his epitaph it would have been that inscribed over the grand master of Alcantara, " Here lies one who never knew fear."

And though Charles V., when he knew of the epitaph, remarked to one of his courtiers that " the good knight could never have tried to snuff a candle with his fingers," yet I hope that no aspersions will be cast on Corydalus by any one who may chance to hear of this second use of the epitaph. For although Corydalus may sometimes have had some fear of something, yet he never betrayed such emotion to me during my short acquaintance with him. And if, in leaving him, I should still reiterate my statement that he is an evil beast, yet, in doing so, I would follow the advice given by Silvius to Phœbe, —

Pupa of *Corydalus cornutus*.

> " Say that you love me not, but say not so
> In bitterness."

CHAPTER XI.

COMPANIONS OF MY SOLITUDE.

> "And every cat and dog,
> And little mouse, every unworthy thing,
> . . . . . may look on her."
>
> *Romeo and Juliet.*

LET me warn the guileless person who has never sought for water-livers. Unless you mean to be a "bug-hunter" all your days, never, never take a dredger in your hand and go forth to dredge. For the one who does that becomes fascinated. There will always be some pool unexplored, some still pond that hides — who knows what? — in its depths. How can one stop when the next lift of the dredger may bring to light some insect-form that one has never seen, but has read of, or found pictured in books? The mysterious is always fascinating, and an unexplored pond is felicity indeed to an entranced dredger. Occasionally one starts up and says, as the last beetle drops into the bottle, "Oh, I must be going home," and then the baleful glitter of a still pool under a willow-tree meets the eye, and one's resolution is forgotten. If Narcissus had known what creatures infinitely more interesting than himself dwelt below the flood, he might not have wasted so much time gazing at his own reflection.

Still, there are drawbacks to one's felicity. I passed by a small house not long ago. A little scrap of a boy stood at the back gate, and, as I vanished down the road, these words were wafted to mine ears, "There's the lady that catches fish."

I am consoled, however, by the appellation, "lady." Let all fish-women be hereafter known as "fish-ladies." To such a pass has come our English tongue.

Let us turn our attention to the brook. But—

"Missis," says a baby voice, and a little boy with a blue apron on looks through the fence.

"What is it?" I answer, knowing well what is coming.

"What you catching?"

And, with a patience born of long endurance, I answer, "Water-beetles."

"Fish for sale. Three for a nickel," remarks a youngster to his friend, as he perceives me passing the oak-tree he is trying to climb.

"Missis, here's a bug," shouts a deceptive youth from the shade of a locust-tree.

But I am not to be deceived into going after that mythical "bug," for I know the "ways that are dark and tricks that are vain" of California boys, and the unbelieving smile with which I look at the group convinces that youth that the brilliant joke is fully understood by "missis."

"Them's *my* weeds," squealed a little skipping girl in the distance, as she once saw me turn aside from the beaten path to look at some weeds by the road where I thought caterpillars might possibly be hiding.

She meant it in fun, just to amuse her companions, for the weeds were not hers, by any means. And the children really had kindly hearts toward the bug-catcher, for it was either the claimer of the weeds or one of her companions who ran across the road, a few days later, as she saw me going by, and cried, "Missis, if I get tins of the red snakes, do you like them?"

On inquiry I thought from her description of "these red snakes that come out of the ground" that they were simply earth-worms, so I was doubtful about accepting the offer; but she went on, "I had two tins of bugs saved up for you a while ago, but you never came by, and so I threw them away."

And this was the little girl that I thought sneered at bug-catching! Bless her kindly little heart!

Most children's talk about bugs comes from curiosity or a love of fun, while some remarks made by older people seem to me to arise from stupidity.

I remember once having visited a brook. It was neither the "other brook" nor this one, but a third, a number of miles from here. I had never before been there collecting, and I found beside the stream a kind of plant with a number of caterpillars feeding on it. Both caterpillars and plant were alike unknown to me, so I caught some of the almost full-grown worms, and pulled up some branches of the plants to take home with me for food for my captives.

I concealed the caterpillars in a box, but the long branches of the plant were not so easily hidden. However, the plant was a very common-place-looking one, there being nothing about it to excite any one's curiosity. But it is ever my doom to fall into the hands of the inquisitive. I suppose that bug-hunters generally meet such questioners, but I hardly expected to be accosted on a city street beside a railroad track.

I had taken the dummy and horse-car back to the city, and I stood near the station waiting for my train, when I was approached by a woman whom I am not aware of ever having seen before, and whom I hope never to meet again.

" What is that you have there ? " asked she, looking at my branches.

" I don't know what it is. Some kind of weed," I answered.

" You ought to know what you 're getting," responded the woman, calmly picking a good-sized leaf from my bundle, and holding the crumpled-up green thing to her nose.

" It is n't anything, is it ? " she went on.

" Why, I suppose it is *something*," I answered, " but I don't know the name."

The woman looked at me.

"Oh, you just picked it to have something green in your hand," she said.

And, perceiving that I had an inquisitive idiot to deal with, I made no answer. But I remarked inwardly afterwards that I had no idea of finding anything quite so green as that woman.

Truly, as the Hávamál of the Elder Edda saith,

> "A better burden
> No man bears on the way
> Than much good sense."

One September day I came to this brook dredging as usual, and found a fat woman and three infants. They were informed that I was catching water-beetles, since they labored under the surprising delusion that it was fish I longed for.

"Water-beetles," said the fat woman with an air of wisdom; "they 're the things that they put on a hook to catch fish with."

No doubt she thought that my fishing was to come. I was merely securing bait. I have become certain that there is one bit of wisdom that all possess. There is no one who is not acquainted with it. It is this: Fish live in brooks.

The general impression that the actions of visitors give to a bug-hunter when beside the brook is that they are inwardly repeating to themselves the sentence addressed by Jaques to Orlando, "I was seeking for a fool when I found you."

And the bug-hunter wishes that it were but proper to respond in the words of Orlando, "He is drowned in the brook: look but in and you shall see him."

Still, I do not know whether the curious gazer would then take the hint and have sense enough to understand the bug-hunter's meaning and to respond as Jaques, "There shall I see mine own figure."

Perhaps, if a bug-hunter could get enough courage, it might be better to say boldly with Jaques, "I thank you for your company, but I had as lief been myself alone."

Once in a while here one hears a soft sound on the bank above, and, turning, sees a red cow peering down among the grass as if to mildly inquire what is going on. But the cow never makes any critical remarks. That is the best of her. She takes it for granted that the rest of the world are employed in sensible business like herself. She and her kin, with that white goat that occasionally looms between me and the sky, appearing, as I look up this little precipice, like a moving day-constellation, a second Capricornus, are almost the only right-minded people that I have ever met when dredging. The goat is fascinating. Did I not in my youth have so violent an admiration for that animal that a certain article of apparel was commonly designated in our house by the humiliating title of "the goat-dress," inasmuch as

I was expected to don that costume every time I went out to play with Billy? Said Billy's perfume being so strong as to infect any article of clothing worn near him.

What mattered it if playful Guillermo did knock me down and with his horns nearly punch the breath out of me? Did I not take revenge on him by drawing his picture one day when he stood cowed beneath a pelting rain, body drawn together, and whole aspect denoting misery? And has not that picture of his disgrace survived until this day? Happy are my memories of that other little goat that learned to play "teeter" with me.

There must be to many right-thinking children one mystery about a goat. It is this. How can he keep his beard looking so smooth, without any comb? Whereas the child who plays with him comes in from the fray with tangled locks a'flying.

The ancient Celts and Cymry had their explanation ready. If there is anything that modern civilization cannot answer, all one has to do is to turn to the simple folk of old, and they are almost certain to have a satisfactory explanation ready. It was the saying of those ancient people that, every Friday night, the fairies made the round and combed the goats' beards to make them decent for Sunday. Truly the *kleine volk* must have a hard time of it if they keep up that custom still. They must be heartily glad when Saturday morning dawns and the job is over.

Poor things, they have had a hard time ever since the first. For the Sclavonic nations say that the origin of elves was sorrowful enough. The first man, they say, had thirty sons and thirty daughters, but, being ashamed of the number of girls, he answered, when asked the number of his children, " Thirty boys and twenty-seven girls."

And, in punishment for his falsehood, three of the most beautiful of the girls were taken from him and changed into elves. Good creatures they were, and they must retain much patience even until now, if during all these years they have been barbers to the goats. What an unhappy trade is that! But perhaps it belongs to the Celtic fairies only.

But there are a few other people who greet dredgers with silent respect. There are some industrious hens who visit here at times. They believe with all their hearts in dredging, for they get many a meal among these water-weeds.

I met a rooster here, too, one afternoon. He was quite tall; at least his shape gave one that impression, and in this respect he reminded me of Mahomet's cock. For believers in the Prophet will remember that he was said to have found in the first heaven a cock of so great size that its crest touched the second heaven, and it is the crowing of this creature that awakes all the animal creation on earth except man. Therefore say the Moslem wise men, " Allah lends a willing ear to him who reads the Koran, to him who prays for pardon, and to the cock whose chant is divine melody." So near together are the Koran and this brook. Art and this brook's visitors might claim some kin, too, for would not these hens be interested in knowing that Gaddo Gaddi at Florence employed himself in making small pictures in mosaic of egg-shells, "finished with incredible industry and patience"?

If men were not so stupidly deaf as not to hear the crowing of Mahomet's cock, perhaps the officer called the "King's Cock-Crower" would never have existed, and so one prince of the House of Hanover would have escaped being startled. For it is written that formerly during Lent the King's Cock-Crower crowed the hour every night within the confines of the palace, and so, on the first Ash Wednesday after the new House came to the throne, the Prince of Wales, afterward George II., was astonished, before his worthy chaplain had said grace at supper, by an officer suddenly entering and crowing the words " Past ten o'clock."

And Prince George in his wrath, not understanding English, concluded that he was being insulted and rose to avenge the wrongs of the House of Brunswick, and had it not been for humble explanations, he would probably have finished the career of the man who tried to be a cock.

I think the rooster at the brook would enjoy

that story. I wish I could tell it to him. He addressed a confidential communication to me when I met him. It was not a crow, but a sort of remark, and I much wished to understand it.

The red cow and her neighboring relatives are helpers of beetles, did they but know it, for in their droppings numerous black scavenger-beetles find refuge, and, on that road yonder, underneath such spots, perhaps one may find in early spring the beautifully marked dark-brown-and-white larvæ of those nimble ground-beetles known as the *Carabidæ*. One must seize such larvæ quickly, or they will retreat into their holes. Thirteen divisions mark the bodies of these larvæ, and to carry so many parts there are six legs and a sort of stumpy, white false leg as a prop at the end of the abdomen. The head looks something like that of the larva of the Water-tiger, being armed with mandibles, while the posterior end of the body finishes with two long hooks. Very eager are these larvæ, when caught, to get away and make for themselves new holes in some clod of earth.

Very few beetles or beetle-larvæ like to be handled, and for that reason I could never more than half believe that tale told of the artist, Buonamico Buffalmacco. I cannot see how he could find beetles that would submit to being treated in the way the tale says his did. He must have been a boy very patient in doing mischief.

For the story goes that he, being when a lad compelled by his master, Andrea Tafi, to rise and paint in the night, devised a scheme to terrify Tafi into allowing him to sleep till morning. Buffalmacco obtained a number of beetles, and fastened little tapers to their backs. During the night he lighted the tapers and sent the beetles through a hole into his master's room.

Tafi awoke, and saw here and there on wall and floor moving specks of fire gleaming in the dark. His superstitious mind was alarmed. Surely these were nothing less than demons. Poor Tafi! He dared not rise to labor, and in the morning the bad boy Buffalmacco duly confirmed him in his belief that demons had visited the room, averring that those individuals hated painters.

"For," said this evil-minded youth, "besides that we always make them most hideous, we think of nothing but painting saints, both men and women, on walls and pictures; which is much worse, since we thereby render men better and more devout, to the great despite of the demons; and for all this the devils being angry with us, and having more power by night than by day, they play these tricks with us."

And so credulous was Tafi that he ceased to rise at night to paint, or to require Buffalmacco to do so. Moreover, the priest was as superstitious as Tafi, and confirmed the notion, and, as every time that the master somewhat recovered from his fright and recommenced painting and making Buffalmacco work at night, he was sure to be punished by seeing the little wandering lights, the priest earnestly advised him to give up the practice. The story became known, and Tafi and the other painters dared not for a long time work at night.

But I should like to see the company of beetles that would allow me to fasten lighted or unlighted tapers to their backs.

Even the geese — genus *Anser*, not *Homo* — that one meets around this brook are prone to look upon me with suspicion. They puff out their fatness and gaze at myself and dredger with an air of superiority that is quite crushing. Evidently, as the Yakimas of Washington Territory affirm, geese were once human beings. Else when did they learn that bug-hunting is foolishness? They seem to be of that opinion now. It was the son of the Sun, so say the Indians, who caused a number of persons to swim through a lake of magic oil that turned them all to animals. But these animals were fat only where they had touched the oil. The person who became the bear dived, and so that animal is fat all over. But the one who became the goose swam on the surface, and that is the reason why geese are fat only to the water-line. No, the backs of the geese will never be very fat, and their scornful glances persuade me that this Indian story is true. Was it not Charles V. who was wont to say of our detested language that one should " use English in speaking to geese " ? — " en ingles á los gansos." But what if " los gansos " will not listen ?

But the scorn of geese has not much effect on a wielder of the dredger.

Down in that grass is hidden a white cat. She is out entomologizing, too, doubtless, and might impart some valuable ideas to a searcher after insects, if she would. But cats are ever averse to giving other people light. The feline race are as stubborn now as they were in the days when poor Tasso, lacking a candle, begged his cat that she would lend him the light of her eyes at night that he might see to write his verses.

" Non avendo candele per iscrivere i suoi versi ! " — *Tasso.*

Freya, the old German divinity of beauty, was said to be drawn in her car by cats. I am afraid that the lady did not travel very fast. Perhaps she preferred a slow gait and chose her coursers in order that people might have time to observe the loveliness of her countenance as she passed.

But woe to the cats when the belief in fair Freya

vanished! For then, of course, people went to the other extreme, and, in contempt of Freya and her coursers, cats with bladders tied to their feet were cruelly thrown from the church-tower at Ypres, as a sign that the people had done with heathen beliefs. Perhaps they had, but that was a strange proof of orthodoxy. I doubt not the pussies would have been better satisfied with the religious rightness of their owners if neither cat-reverence nor cat-contempt had prevailed.

Still I do not know that the custom of Ypres was any more cruel than the ancient Scottish method of treating cats. For, anciently, when some one of the tartan-wearers wanted to know something of the future, he would catch a live cat and hang it up before the fire, and leave it there until by its cries it had attracted other cats to the place. Then the worthy Highlander would ask the questions he wished to have answered, and would interpret the cat-cries so as to make replies in the Gaëlic tongue. And then, when the Highlander was fully satisfied, he had mercy on the singed cat and set it free.

Even Paris was not kind to cats, for, on the eve of St. John, the mayor of the city used to put about a dozen cats into a large basket, and then, kind man, proceed to throw the unlucky beings into the bonfires that were built on that festival.

All this would be horrible news to that cat down in the grass. I am afraid she would faint if I told it to her. One must be careful about telling evil tidings too suddenly.

That cat would feel much more flattered to hear of the "Cat's Fugue" that Dominico Scarlatti made from those five notes that his cat struck when she jumped on the piano keys. If this cat could understand, what an interesting tale could one tell her of that painter of Switzerland, Gottfried Mind, who was called the "Raphael of Cats" because he made so many pictures of the feline beauties. And how sincerely would this cat sympathize with Gottfried's grief, when, on account of some signs of madness that had shown themselves in the cats of Berne, the magistrates of the town gave orders that the cats should be killed! Eight hundred cats are said to have perished in that fearful massacre. Imagine what Gottfried's feelings must have been! How terribly lacerated! Indeed it is said that although he managed to save his own pet cat Minette from this St. Bartholomew of the pussies, yet he never afterward seemed to be wholly comforted.

But, the killing-time being over, he set about painting pictures of more cats and kittens than ever, and the next winter he went into the business of cutting chestnuts into the shape of his adored pets. And so "cute" were his chestnut-cats that people bought them as fast as he could make them. I fear there will never arise another "Raphael of Cats."

"Il cane e fedele si, ma il gatto è traditore." Dogs are faithful, cats are traitors, is the Italian sentiment. But I had rather meet a cat than a dog when I am out after "bugs." There is a dog that lives across the creek. He is a despicable person with a mighty voice, and his sole desire is to break loose and rush over and bite me whenever he sees me looking among the branches of that weeping willow. He takes it for granted that I am in bad business, and I naturally feel aggrieved. Perhaps he is color-blind, and mistakes my black dress for the blue blouse of a Chinese rag-picker, for I have seen one of those persons here searching this little ravine for additions to his baskets.

No cat ever has spoken so impudently to me when on these entomologizing excursions as that dog has. He is a villain, and a standing argument to me that the Ettrick Shepherd was entirely wrong when he said, "I canna but believe that dowgs hae sowls."

As a general thing, however, the companionship of "critters" is consoling. Go out some day when you are disgusted with your neighbors, when the inquisitiveness and idiocy of the world of human beings has been particularly galling. Go and sit down on the green grass by a cow for a while. Listen to her speech, look into her eyes, observe the soothing regularity of her chewing. See with what confiding trust the goat approaches you. Pat him on the head and break branches of the oak for him, and observe with what bliss you have filled his soul. And, after an hour of such treatment, see if the "blues" have not departed. Whither? Upward, perhaps, to form part of the sky. Of course there may be drawbacks to this treatment. The grass may be damp, and you may get rheumatism, but that is your own fault. The axiom I set out to state is that in cases of "nerves," the companionship of animals is a sort of anodyne.

They look up at you with a sort of humbleness or audacity as suits their respective personalities. And when two black kids dance for you the cachucha, or whatever else you may call that extraordinarily graceful exercise that young goats caper through, you feel mightily pleased. Thor himself, you reflect, was drawn in a chariot by two goats, Tanngnjost, the "teeth-gnasher," and Tanngrisner, "fire-flashing teeth." You hardly attempt to pronounce the names of those animals, even to yourself, but you wonder if in that Scandinavian myth there does not lurk the recognition of the fact that the grandest personages may find comfort in the society of animals. For that matter, lower creatures than animals can bring consoling diversion to the mind, for reptiles and insects have powers of amusing unsuspected by those who have never watched them.

CHAPTER XII.

FROGS, BOYS, AND OTHER SMALL DEER.

"Come hither, boy."
Titus Andronicus.

On the higher ground and the hills near this brook dwell other creatures. In that garden yonder I have often seen the great, yellow, slimy slug, Ariolimax, so frequently found in damp spots along the Pacific coast. I am afraid I stocked that garden with those slugs. Being up among the foot-hills north of here I found several of the slugs, one measuring about four and three eighths inches long. Six or seven inches is a length attained by these creatures sometimes. I captured a big one and took it away with me.

Little did the people in a certain restaurant that evening think that a prim and proper person sitting at one of the tables, calmly partaking of her supper, had carefully tied up a huge slug from the gaze of the world and now had the creature waiting beside her. But it was even so. I brought the slug home. I think that was about two years ago, and now in the twilight of April evenings, or even occasionally in a September rain, various Ariolimaxes drag their bulk into sight upon the walk. One evening a little girl called my attention to a slug that was about five inches long, an inch and a half high, and about as wide. The slug looked a good deal like a slimy, glistening sweet potato. Even the infants among these slugs are of greater size than ordinary. I frequently see partially grown slugs that measure perhaps three inches long, but are only about half an inch wide.

Ariolimax has a dainty taste. Blue violets are delightful articles of food to these slugs. I remember how horrified two ladies were to meet a moderate-sized Ariolimax among the leaves of the violets they were picking. The slime left by such large slugs is very tenacious. Get some of it on your hand and it is a difficult job to wash the substance off. Step in it, and you will wish you had n't. The promenades that the Ariolimax family insist on taking toward dark become a serious source of annoyance and worry to one who does not wish to set his foot on one of these huge creatures. I should as soon think of crushing a baby.

I once found one of these huge slugs inside of General Fremont. For the benefit of the too startled reader I would hasten to explain that the General mentioned is a huge redwood, one of the "Big Trees" near Santa Cruz. Damp woods and the vicinity of springs delight Ariolimax and in such places this creature may hide during the dry summers.

What varied tastes there are among these higher ground creatures! Some of them are the opposite of Ariolimax in their likings.

On dusty paths over these hills in August, you will sometimes see, by looking sharply, a little creature skipping along by your feet. If you are not on the alert you may mistake him for a bit of dry stick or straw blown by some minute breeze. Catch him if you can and put him under a microscope. He is one of the Thysanura, or Springtails, a narrow creature, with long antennæ in front, six legs in the middle, and some bristles behind. One that I caught could jump in a very lively manner. He seemed to be covered with butterfly-like gray scales that came off from the poor fellow when I turned him over on his back or in other positions to see him plainly. He had, I think, six-jointed antennæ, and three anal bristles.

Wingless creatures are the Thysanura, but what does it matter? Perhaps hopping is just as much

One of the Thysanura.

fun as flying to one who is used to the former method of progression.

There flies a bee. Job was responsible for the existence of bees, I believe, according to the rabbins. They anciently told a grisly tale that before the days of Job there were neither silk-worms nor honey-bees upon the earth. But, after that righteous man's great affliction and his restoration to prosperity, the worms that had devoured the body of Job were turned into silk-worms, and the flies that had tormented him were changed into honey-bees.

Mahomet said that bees should be admitted into Paradise, although I fear the gentleman was about as far wrong in his assertion as the rabbins were in theirs. It is at least a queer thing that there should have once been thought to be so intimate a relation between bees and souls. I believe it is said that in some parts of England bees are not allowed to leave their hives on Friday, on account of the religious scruples of their owners. And the French peasant women go on the Day of Purification to read the Gospel to the bees. How very wicked must wild bees that have no religious privileges appear to such people.

These *Hydrometridæ* are very numerous here. You will see different sizes of them. In the daytime, I mean. The Skaters go to bed at night, like other respectable citizens.

I discovered this fact by keeping some of these creatures in a miniature pond. At night they went to bed on the sides of the lake, on the earth above the water. Neither were they dissipated creatures, for I found one of the Skaters going to bed as early as a quarter before six on a June evening before it was dark.

They did not like to get up very early in the morning. I have gone to the miniature lake at half-past six in the morning, but the folks were not up. By looking sharply, I could see one hanging above the water, his hind legs flopping helplessly down below him. In a bigger hole were one or two others, as could be seen by the number of legs protruding from the cranny.

But the folks were not at all disturbed by my visit to their chamber. Perhaps they were dreaming. Did they dream of the brook where they once skated as free bugs, or did they dream of a time when their jailer would be able to catch more palatable flies for them than those were that were thrown in yesterday? Again, at half-past seven, I looked in upon them, but the sluggards were still abed. Water-skaters are evidently lazy.

And I do not think that they have as quick eyes as one might imagine them to possess. For, in idle moments, when I have stood beside some little pool on which the Water-skaters were numerous, after the first shock of my presence was over and the creatures had forgotten me, I have amused myself with casting broken heads of grasses into the water, as near the creatures as I could, to see if they would mistake them for moths or other living morsels falling within reach. And, although many times my grasses were unnoticed, yet, at other times, one of the Skaters would turn hastily toward the grass, or steer himself to it, to make sure it were something edible.

Perhaps, however, such a Skater may have hoped to find some bug clinging to the grass, but the action of the moment seemed to indicate that the Skater had been deceived. Yet the Skaters are very quick to see a person approach the pool, and after sitting on the bank awhile one has only to stand up to send the whole company of Skaters fleeing away over the water.

A boy once told me that these creatures have fights, one that has something to eat being chased around by several others. I do not know whether the Water-skaters ever attempt to capture the Whirligig beetles. I have seen a Whirligig pass directly before a middle-sized Skater, and yet the latter did not seem to care for him. Water-skaters and Whirligigs can be seen in joint possession of the surface of some small pond. Perhaps the Whirligig is too hard a morsel for the Skater to enjoy. In the few experiments that I have made with Water-skaters on the surface of pools, I have not found the insects liable to be deceived like the Whirligigs with bits of red, white, or dark cloth. But I have held these by white strings, and perhaps the Skaters saw through the device on that account. When I deceived the Whirligigs they were in captivity, and the thread with which I let down such bits of cloth may have been so fine as not to be noticed.

One finds in May some of the little Water-skaters, very minute as compared with the adults, yet looking like them as real bugs generally do. The little Skaters are very nimble, and seem to know immediately that a bottle is not their customary home. So small are they that unless you are near a pool, looking intently at the surface, you would hardly notice what they are, and I do not think that it enters into the boys' heads that these are the same as the big Water-spiders. At least one boy that I spoke to about it seemed to be ignorant of the fact. But such minutiæ of pond life is not generally regarded by boys. I have heard one of them call a snail a "beetle" and another call a water-shrimp a "bug."

The latter fresh-water crustaceans are very numerous in these pools. The sticks and hiding-places are alive with them. They have five joints to their legs and about thirteen segments to their bodies. Probably they are akin to *Gammarus robustus*. They are a nuisance, for the bug-hunter can hardly shake the dredger clear of them. Some will stick until the bug-hunter is on the meadow going home, and will then crawl out and demand to be taken back to the brook. One hastens to shake the crawling things out in the midst of the grass and leave them to walk back to the pool. If they were beetles with wings, one could easily drop them and relieve one's conscience by the remembrance that they could fly. But such mode of progression is denied the water-shrimps, and if it happens to be a day when the bug-hunter is specially tender-hearted he will turn back and shake the dredger over the brook once more. If it be a hard-hearted period, the bug-catcher proceeds onward with a guilty consciousness that he will see the dried bodies of those shrimps adorning his dredger next day. Whichever way the bug-catcher decides, his peace of mind is destroyed, and he mentally anathematizes the water-shrimp.

Here on our pathway, as we go to get "bugs," is an old shoe, cast away on the hill-side. Was there ever a fabled divinity of the ancient time to whom old shoes were sacred? Memory saith not, but to Vidar the Silent, the son of Odin, were due the scraps of leather that were cut from the toes and heels in making patterns for shoes, and the Norse shoemaker who wished to assist the gods was charged to throw away all such pieces, since it was supposed that they went to make Vidar the Silent's shoe. Of this it was told, "It is a thick shoe, of which it is said that material has been gathered for it through all ages." Perhaps, if Vidar had not been too proud to receive offerings of old shoes, he might have constructed his own foot-covering the sooner.

Pop! pop! pop! in a succession of splashes, go the frogs into the water, till the sound resembles a small cannonading extending up the stream in honor of one's coming. Small idea have the frogs of welcoming any one, however. Sunning themselves on the banks, they hear the sound of footsteps, and hastily arise and throw themselves beneath the flood. Good reason, too, have they for such cautiousness. One boy informed me that he made sometimes seventy-five cents per day in catching frogs for a Frenchman, and on a remarkable day this boy earned a dollar. No wonder the frogs flee.

There is something pleasantly meditative about a frog that sits on a bank. One feels inclined to sit down beside him and inquire the subject of his meditations. Why should the heralds of the Middle Ages have used the frog as a symbol of degradation? I would not reveal such a fact as that to froggy, if I sat beside him. I would, instead, point out to him the blessedness of the fact that there are no pike in this brook, and quote to him Izaak Walton's words: "It is observed that the pike will eat venomous things, as some kinds of frogs are, and yet live without being harmed by them; for, as some say, he has in him a natural balsam or antidote against all poison." Did not Queen Elizabeth call Francis of Anjou her "Frog"? It was not complimentary to Francis' looks, but froggy would not understand the sarcasm.

I have my own idea why *Vanessa Antiopa* wears black. I also would wear mourning if I flitted daily by this brook and saw the cruel deeds done here. One day in April when Vanessa was winging up and down the creek, I was passing under the willows where their dropped-off blossoms lay like so many fuzzy, brownish caterpillars on the grass, and I made my way to a well-known pool. Horrors were enacting there.

Vanessa Antiopa. Mourning Cloak.

Is there anything more cool than the way in which human beings assert their ownership of smaller beings? "Here's *my* frog," cried a little fellow as I drew near. The youngsters were engaged in the time-honored custom of killing a frog. The poor creature, already hurt, was swimming with the little strength left it, and a boy caught it up only to throw it back into the water as a target for more shots. I fled from the spot.

I wonder that Vanessa can bear to spend her life by a creek that has such sights to show, but I think she shows good taste in dressing in mourning.

If I am ever the editor of a newspaper I shall certainly write an editorial on "The Opinions Of Frogs." "*Qu'en disent les grenouilles?*" "What will the frogs say?" was the old court-phrase at Versailles when high-born beings spoke of the inhabitants of Paris. Would that a little more deference to the opinions of frogs might prevail nowadays.

Bah! What becomes of our boasted civilization when it is looked at through the eyes of one of the lower creatures? Have I not seen a person, otherwise intelligent and good-natured, slap ants so as not to kill them, but leave them painfully stumbling along wondering what right such a giant had to put them in pain? Have I not known a person to give partially stunned flies to Skaters? Have I not known a teacher to dissect a sea-urchin *alive* before his class? As for those abominable "naturalists" who pin bugs, beetles, and flies alive and leave them to die lingering deaths, I have no words strong enough to say of them; and if I thought that by writing this book I should interest some person who would catch and torment any of the lower creatures, I would put my manuscript in the fire. Better is it with the Moslems to let a moth fly by unmolested, believing that it is a " messenger sent from the dead to see what is transpiring upon the face of the earth," than to have no more tenderness for the creature than to pierce its vitals with a pin and leave it to endure agony.

But there is small use in lecturing boys. Another day I came across two of them trying their slings on a polliwog in a pool. They "smashed" him, as one of them stated, and, on my expressing a wish to know why they did such deeds, one of the boys informed me that the polliwog had no business to be there.

"Why, yes, he has," I said. "He was hatched here."

"How do you know where he was hatched?" queried one youngster.

"Well, it was in this brook," I responded.

And then, proceeding in a severely virtuous vein, I said, "*You* would n't like it, if you were a polliwog."

"Yes, I should," said the Incorrigible. "I'd sneak away where they could n't find me."

"Maybe you would n't have sense enough to do that," I answered, preparing to leave such a reasoner.

"Yes, I should," answered he, with perfect faith in his own powers.

And, as I climbed the bank, he sent after me this awful threat, "If *you*'s a polliwog, we'd smash *you*, too."

Would he, indeed! Depravity, thy name is Boy.

And yet my conscience smites me when I say so, for have I not held entertaining conversations with boys along this creek, and have they not told me things about water-creatures that interested me much? Sometimes the things were true, sometimes false, but, at least, the boys believed them, and did their best to impart to me their knowledge. Let me not be ungrateful.

Beside this pool one morning came a youth (I think the same one who made the fearful threat about the polliwog), and lamented to me that he had been compelled to sell two frogs for a nickel apiece.

The subject of frogs was called to our minds by the sight of one dead in the pool. The boy confessed that he had killed that frog, but he said it was too nearly dead before for him to sell it.

"Ought to have had a dime," said he, referring to the sale he made to the Frenchman. "*I* would n't eat them for a hundred dollars. The Frenchman says his brother eats them. He says they 're nicer than any other meat."

And it seemed to me that two nickels was rather small pay for the labor of splashing around in a pool catching two frogs, and then tramping several miles to sell them. However the boy was not discouraged. The pool had been clear when he caught the two, and he thought he had seen two others in it. He was coming back after them.

And then ensued a talk on the prices of frogs.

Some boys must have a faculty for getting more money than others out of that Frenchman, or else the frogs they catch are bigger than those caught by others, for this boy, on my mentioning one fellow who had received a dollar, said, "That's for a dozen."

The dead frog in the pool was estimated by this young financier to be worth about ten cents, if the creature had been in good condition.

This youngster and his companions tramped on up the stream, and left me to sit down on the bank in blissful solitude. But it was only for a moment. A boy and two dogs came along the side of the bank. The boy was bent on giving the dogs a bath, and he did so, seizing them by the neck and back and throwing them in. They were good sized-fellows, half Newfoundland, but not yet full grown, and they did hate their bath, but they had to endure it. Their master regaled me with a tale of the way in which he had tried to make them swim in salt water, and how they swam in it to some schooner, thinking it was land, and then the poor, disappointed brutes had to turn around and swim back. The boy was afraid they would drown, and indeed they were quite exhausted on reaching land.

This boy was quite a communicative, pleasant little fellow. He was full of ideas about bugs, and anxious to know more, and volunteered the information that frogs have to come to the top of the water to breathe. Finding that I agreed with him in this, he furthermore informed me that my dredger was a fine one. It was rather amusing to hear him admire the ragged old thing, but I believe boys have coveted it before. He said, though, that it was not the kind to catch frogs with; they would jump out; one must have a long, sack-like dredger. And he said that frogs have regular sleeping-places in holes on the sides of pools.

"You can come to this pool at night and see them," he said; "some go in the grass, and the same frogs go in the same holes. They keep their heads a little out of water all night."

I unwittingly informed this boy of the intentions of the former one in regard to coming back after those two frogs. My second visitor's cupidity was aroused, and he announced that, if he could get one, the other fellow would be disappointed.

This second boy proceeded to describe to me the water-snakes found in the other creek.

"There are none in here," said he; "I've hunted for them."

And I could verify that statement, for I have never seen any here.

"They're about as long as your stick," he went on, referring to the nearly yard-long handle of my dredger. "They've got a red streak on top of their backs, and they're gray, with black streaks on the sides. They bite, and they're poison. The boys catch them with a noose of grass. Some of those oats would do."

And springing across the pool, the boy picked a shoot of oats, stripped off the heads, and proceeded to make a dainty little slip-noose of the thin end.

"You hold that in the water and let the water-snake go through it, and then you just draw it up and catch him," said he, suiting the action to the word; "only you must draw it up just back of the head or else they'll bite you. We catch lizards that way."

"Water-lizards?" I asked.

"No; just any kind of lizards," said he.

"What do you do with the water-snakes?" I asked.

"My father sold three for me the other day," said the boy, and then he mentioned the name of a San Francisco druggist as the purchaser.

"What does he do with them?" I asked.

"He's a snake-charmer," said the boy.

And, seeing that he would answer any question that I had a mind to ask him, I queried, inquisitively, "How much did he pay you for the water-snakes?"

"Half a dollar apiece," said the boy.

As I sat beside the pool I twice saw a leech swimming through the water, its body undulating as it moved.

The boy caught sight of it and said, "That's just the way the water-snakes swim."

"They go down differently from what they come up,— don't they?" he went on, as the leech descended near the further bank; "they go straight down, and they wiggle when they come up."

Maybe so, but I never noticed it.

My talkative companion furthermore stated that "up by the college some boys found a gopher-snake with a weasel in its mouth."

Ah, many and many an interesting thing have the boys told me. It has been part of my craft to worm information out of them. They are so ready to tell all they know and are so often correct in their observations. And shall my pen revile my benefactors? But one aches to shake them when they are cruel.

Nor is cruelty limited by age, for did I not find a five-year-old and a perhaps four-year old — mayhap only three — on the top of that bank, engaged in tormenting a small butterfly and a moth? The boys would toss their prizes into the air, let them fly a little way, and catch them again. I remonstrated, but in vain. The younger one opened his hand to show me his moth, and I was invited to have a "lady-bug," as one of them called a *Diabrotica* that he had found. One boy announced that they were going to catch grasshoppers, and I think that the little butterfly was almost dead when I went away.

I herewith confess that I stole that younger

boy's moth. I should not have deprived him of it if he had behaved himself, but he went too far. He had done enough in keeping it shut up in his dirty, hot little hand.

But, by the time I had walked down the road to a pool, the boys came rushing toward it, and before I knew what he was about to do, that younger imp threw his moth into the water. The poor thing struggled, and calling some oats to my assistance I fished it out and clapped it in my tin. Moreover I delivered a small-sized sermon to the imp.

"Give it to me," said the former owner.

"You can't have it, if you're going to be so naughty."

"Give it to me," sternly demanded the boy.

I looked at him. He had run out-doors so much without his hat that the skin looked almost ready to peel off his forehead.

"Give it to me," reiterated he, with all the impressiveness at his command.

But I turned and climbed the bank while he proclaimed his desire to throw that moth away out there on the water, and the youngsters announced that they were going to catch some more things and throw them in. But I think that was only an assertion of helpless wrath, for I passed by the pool a little while after and no such work as had been threatened was going on. Privately I was glad that the boy had thrown the moth into the water, for it gave me a chance to take the creature home and look at it, a thing I had wanted to do from the first. I had an idea that my booty might possibly be *Arctia virgo*, the Virgin Tiger-moth. But it was not. It had pectinated antennæ, yellow fore wings marked with rectangular blocks of black, its abdomen marked in the same style, while on the hind wings was a tinge of red, in addition to the dark marks. Having observed so much I let the poor thing go free.

"I wept when I was born, and every day shows why," says old George Herbert. Perhaps moths also groan to themselves about the bitterness of life. If insects could express their feelings toward their human persecutors, I think sometimes the creatures would join in something like the old French song : —

> "Ah, ça ira, ça ira, ça ira,
> Les aristocrates à la lanterne."

I am afraid that the attractions of this brook prove too great for the integrity of the boys at times.

"Somebody's playing hookey," I heard one boy say, as he came up to a group of three others under some willows.

"So 're you."

"I 'm not."

And the accused boys both repudiated the suggestion with virtuous scorn, although a minute afterward I heard the school-bell ring and no one of the four started. They were looking at one of their number who, as far as I could judge, had a little while before this fallen into the brook and was in the process of being restored to dryness, his jacket being spread out on the grass. "Nature's unhoused lyceum" had more charms for the four than that school-house on the hill, and they passed me afterwards, bent on some errand by the brookside, yet stopping to inquire if I had found anything; meaning bugs, of course.

I have one fear. It is this. I came up a bank a while ago, and in the road above it were a number of school children playing. As I passed with my dredger covered with paper, and my spoon and bottle held unostentatiously in my hand, I heard in the midst of the noise and confusion what sounded like an indistinct remark about "frogs." There seemed to be no appropriateness in such a reference except that I was near. Now I can endure with equanimity being called a "fish-lady." That is an honorable business. But, if I am about to be looked upon as a "lady that catches frogs," I demur. My heart begins to fail me at the prospect.

The boys do not suspect me of such a thing, or they might not tell me of their trades so freely as they do. How wicked should I be to take advantage of them and steal their business!

But this Frenchman and his brother are far from being like "poor Tom that eats the swimming frog, the toad, the **tadpole, the** wall-newt, and the water." The Frenchmen are more particular, and if a frog is **not a " red-legger "** it is not salable to them.

So are the **boys** driven **to** using their eyes in picking out their wares, for it would be a sorry thing indeed to carry the **wrong** objects several miles and then be met with a flat denial of money in return for them.

It is said that the old, bald, wrinkled Paul of Russia, who was so ugly that he did not dare put his countenance **on** his coin, issued a proclamation prohibiting, under penalty of killing by the knout, any one of his subjects from making use of the expression " bald," **in** speaking **of** the head, or " snubbed," in referring **to** the nose. And, moreover, the same gentleman, with the characteristic fear of the Czars, forbade the academy to use the word " revolution " **in** speaking of the courses **of the stars.**

I suppose there **are** tabooed subjects **of** conversation with all persons, and I should think that the special objects to be avoided **in** conversing with frogs would be " boys " and " Frenchmen."

A little Portuguese **boy** that I found several days after my conversation with the boy who had the dogs gave me a much more moderate estimate **of the** length of water-snakes than the former boy had given. The little Portuguese said that the snakes went " down in the mud."

But I came to the conclusion that perhaps he hardly knew the creatures after all. His face expressed a doubt of himself, although he asserted that the snakes were in the other creek. He had seen them there. Another boy with him volunteered the guess that they caught "fish and things."

In like manner did I hear a vague rumor from one boy of a kind of bug, in the other creek, that "had horns behind." Taking the peculiar description that the boy did his best to give me, I should judge that he meant the kind of Water-scorpions that bear breathing-tubes behind like Ranatra. There are no such Scorpions in this brook, I believe. In all my dredging I have never found any of that tube-bearing variety, *Nepa*, in this stream. Probably, from the peculiar manner in which the scorpions of this brook carry their eggs, the creatures belong to the genus *Belostoma*.

There is a great clump of the white-veined thistle opposite the willows. There in a thistle-leaf I once found the caterpillar of the butterfly known as the "Painted Lady," *Pyrameis cardui*. The Lady preferred living alone, as all her folks do. In one spiny leaf she had made a little web, and an intruder had to break into her home to examine her, since she had drawn the two sides of the leaf partly together. I bore the Lady home, and heroically pricked my fingers many a time in obtaining her food. She did not seem very friendly. I do not see how any one who fed on such very spiny thistles could be.

But she grew at a startling rate, and, one June day, relieved me from making any more trips to thistle-bushes by turning herself into an angular brown chrysalis, adorned with golden tubercles. She lived in this style while I packed her up and took her with me on a journey of a hundred and twenty-five miles. Then, one July day, the joints of my Lady's chrysalis began to look juicy, and great wriggling took place. My Painted Lady came out of retirement gorgeous in coloring. One evening, taking my Lady, I walked out into the forest, and having found Achilles in the shape of his namesake the white yarrow, *Achillea*, I laid my Lady at the feet of the gallant Grecian for protection. My Lady clung to my finger as though loath to part from me, but she was soon made to understand that separation was inevitable, and she subsided under the yarrow leaves. There, as she held up her wings, all her brilliant colors were hidden, and the gray under-surface was so much like the general gray shade of the yarrow leaves and the grasses that I could but just distinguish her as I stood up. So she nestled down at Achilles' feet for the night, and I saw her no more. Perhaps, in her flittings through the pine-woods of that hamlet by the sea, she has ere this found her destined mate, a butterfly that would never have had a chance to see my Lady had I not been the cause.

Another companion that I took on my journey was a female "Tussock Moth," *Orgyia*. She had no wings, as the lady-moths of her variety do not fly. One can find the *Orgyia* caterpillars in this district, though not in great numbers. They are pretty creatures, small, with four gray tufts

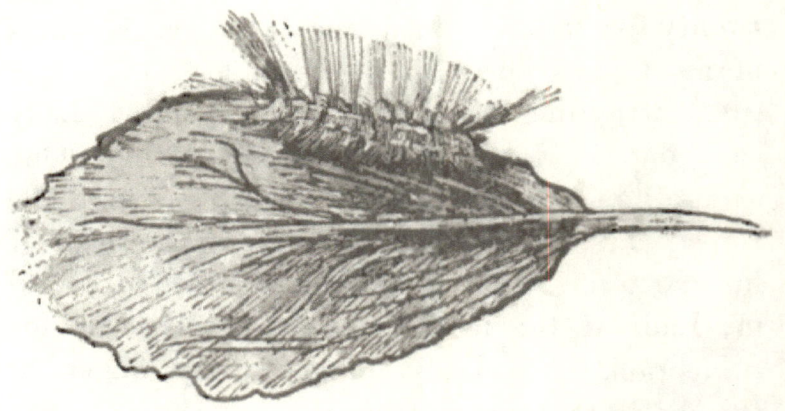

Caterpillar of *Orgyia*.

on their backs like a camel's hump, and the gray hair hangs over their heads like bangs. The segments are marked with red and yellow, but the most conspicuous things about these caterpillars are the long tufts of black hairs. One stands out on either side of a caterpillar's head like antennæ, and another such shoot adorns the end of his body.

I fed mine on apple-leaves, and it was on such a leaf that one made a thin, fuzzy, light-grayish cocoon. A winged male with the comb-like antennæ of the *Orgyia*, and grayish wings, brown underneath, made a hole in one end of his cocoon

Cocoon of *Orgyia*.

so neatly that a person looking at the cocoon would not have known that the moth had come out.

It seems strange to see a moth without wings. These lady-moths lead very stupid lives, hardly

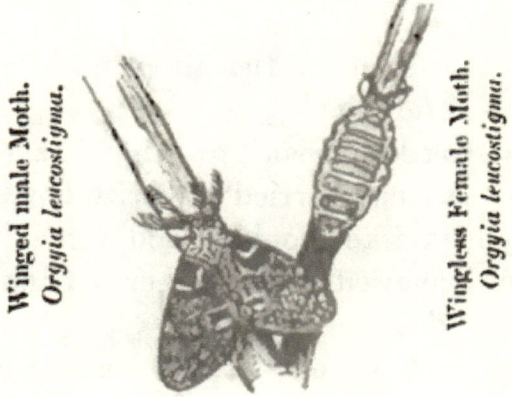

stirring from their places during their whole existence, sometimes, and consequently it was a very startling event for Madame Orgyia to be transported to a different section of the country. But she did not object in the least. She looked somewhat like a beetle that had lost his intellect. She gave but little sign of intelligence, but, while away with me, she laid six small eggs, that looked like young pills.

One day as I wandered beside this brook, bent on ascertaining what manner of creatures lived on its borders, I came to a tree, and, peering in among its leaves, I spied something that gave me to un-

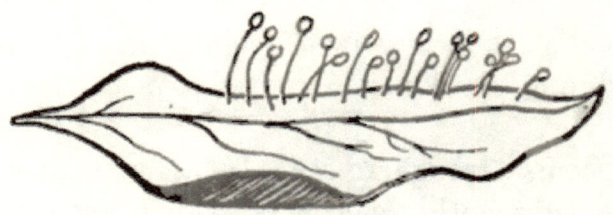

The leaf I found.
Eggs of *Chrysopa*. Golden-eyed Fly.

derstand that some of the Brookside People had been there before me.

"They look like beans, growing," thought I to myself, after I had carried my prize home.

"They look like mould," said a girl friend to me, when I showed them to her. I thought we were both right.

On the back of one of the leaves were what looked to my friend like a fungous growth, but to me like twenty little beans not bigger than pins' heads. These green "beans" were all standing up in the air on five-sixteenth-of-an-inch long stems, much as real beans hoist themselves up when they begin to grow. Unfortunately I broke the "bean" off one of the stems, so but nineteen remained. The "beans" were oval, opaque, and inclined to shiver on their slight stems like a number of reeds.

I took the leaf home, and looking at it, I saw at the bottom of one of the bean-stalks a little green plant-louse, or aphis. He was standing looking up the stalk as if he longed to climb. It was quite a tree to him. He did not know what enemy lived on it. If he had climbed, it would have been almost a repetition of the story of Jack and his Bean-stalk, except that perhaps the giant might not have been awake by the time Aphis arrived at the top.

For, although little Aphis did not know it, there really was a giant at the top of the stalk, as there was on Jack's, a giant that was coming down by and by to startle all the Aphides.

The "beans" were really eggs, and one day, after they had turned to a lilac color, they hatched. How those Giants of the Bean-stalks ever got down from such heights I do not know. Perhaps they tumbled. If they did, it must have been like falling over a precipice to them.

But when I looked, there they were on the leaf, queer little mites of brown things, with six legs and a pair of nippers apiece. The eggs that the

Chrysopa larva.
One of the "Giants of the Bean-stalks" (A little larger than mine when full-grown.)

take my eye away and see the monster dwindle to a dot. He was a very emphatic dot, however. A whole world of determination was in him.

Blood-thirsty Giants these creatures were for one-day-olds. I hunted some Aphides for them, and the Giants went to work at once. It is "excellent to have a giant's strength," but more especially to have his jaws in this instance. No matter if an aphis were almost twice as large as a Giant, into that aphis' side went those dreadful little pincers, and the Giant held on.

Wise men tell us that the mandibles of these larvæ are hollow underneath, and that the little maxillæ exactly fit into these grooves, and so make a pair of tubular forceps through which the juice passes from the aphis into its eater. Wise men tell us another thing, too, and that is that the feet of the Giants are particularly fitted to let them run over the twigs and leaves of plants safely so as to find the Aphides. In Europe the gardeners hunt for the Giants, or, as they are more usually called the "Aphis-Lions," and put them on trees that are overrun with Aphides. No doubt the Giants are delighted with such attention. The trees are soon cleaned if enough of the Aphis-Lions are found.

It was necessary to paste paper over the top of the jelly-glass in which I kept the Giants, or else their lively food might have run away. But it became quite a task to hunt enough aphides for the Giants. Their appetites were very good. Rose-bush after rose-bush did I search, and sometimes I found what I wanted on wild mallow plants. The Giants grew finely, and each developed a red line on his back.

One day I gave my Giants an extra good dinner. I had found a little rose-bush in one corner of the yard that I had forgotten, and it was covered on its fresh ends with hundreds of Aphides. I captured a great number and put them in with the Giants.

Next day I looked to see if more were needed, but I could see plenty through the glass. The next day things were the same, and the next day after that I began to wonder what was the matter with my Giants' appetites. So I poked a hole in the paper and opened the jelly-glass. The liberated Aphides came out rejoicing, but where were the Giants?

I began to pull the dry rose-leaves apart, and lo! here and there were little white balls that looked like pills, only they were somewhat cottony on the outside. The Giants, being of the same age, had all become sleepy at the same time, and passed into the pupa stage. The cotton of their

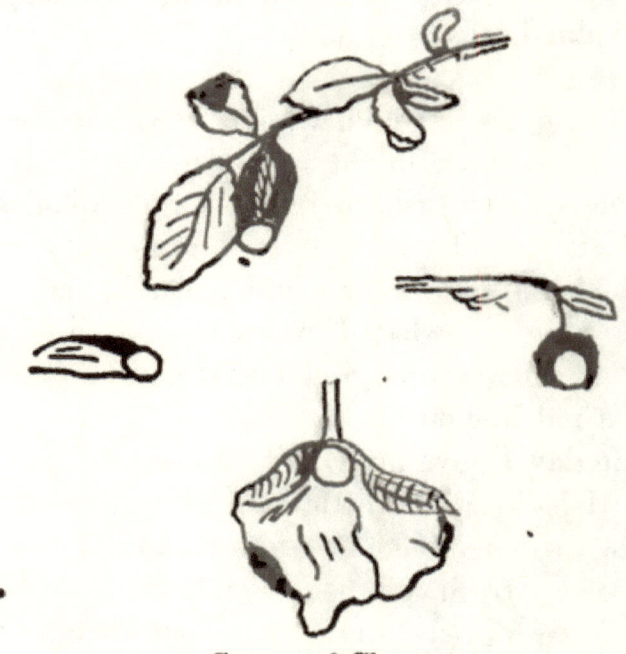

Cocoons of *Chrysopa*.
Mine were just a little larger than this.

cocoons was beautifully white, and one could imagine the Giants sleeping quite comfortably inside such sheets. I rejoiced that the last meal I had given them was so good a one, for I should have felt guilty if I had allowed them to go to bed hungry. If Aphides notice anything of the sort, how they must have rejoiced when they saw the Giants becoming too sleepy to eat them!

There were only seven of my cocoons, for the number of the Giants had been mysteriously diminishing for quite a while. I suspect that they were not always as good to one another as Giants should be, and I am afraid that they perhaps hurt one another so much with those scissory forceps that some of the Giants died. I know I found two that I thought were fighting, and one seemed to be injured by the contest. Although I did not see any cannibalism, yet I know of no reason why the Giants might not have tasted well to one another, since they must have been composed inwardly of juice extracted from the Aphides, and so perhaps have had a similar flavor.

The usual length of time of the pupa state with the creatures is, I believe, fifteen days. Mine, being the winter brood, slept longer. For five months, commencing with about the 10th of October, the Giants slept in their cottony sheets in the jelly-glass. I put it on a shelf in a dark closet, and the 19th of March I looked to see what was coming to pass there.

I found that four Giants had awaked. Their cocoons and husks were lying on the bottom of the jelly-glass. One Giant, alas! had waked but to die. I found his body. I suppose that he died during transformation, as so many creatures do. Several Giants that were sleepier than the others afterwards rubbed their eyes open and arose.

But the Giants were not as of old. They had taken to themselves beautiful gauzy wings, and were pretty creatures with green bodies and bright yellow eyes. *Chrysopa*, the Golden-eyed Fly, is the rightful name of these creatures, albeit to me the eyes look very much as though made of brass instead of gold. But I would not insult *Chrysopa* with such a suggestion. Moreover, *Chrysopa* is not a fly at all, but belongs to the *Neuroptera*, being a distant relative of the Dragon-flies. Perhaps, on some leaf that I shall never see, there have stood ere now some more little "beans" set up by one of my Giants, and the story of the bean-stalks may all have begun over again.

Chrysopa.
The Golden-eyed Fly.

CHAPTER XIII.

A LINGERING GOOD-BY.

"And these things finish."

Shakespeare.

It is time to close. And yet there are many things not set down in black and white. People might learn much beside this brook. One marvels that they do not come here to study. But I reflect on Peter the Great and am comforted. People are not far different from those of his times. Poor Peter! After establishing at great expense that large museum of Natural History at St. Petersburg, he was driven to offering to his beloved subjects a glass of brandy apiece as an attraction that would draw them to look at the wonders of creation in his museum. No wonder this open-air museum of to-day does not draw people. Nature offers no stimulant here, save the golden one of sunshine, tonic enough for those who count their descent from Father Adam.

But stay. If one is thirsty, one might take a dock-leaf. Does not the fine name of the dock, *Rumex,* come from an old Latin word meaning "to suck," since the Romans when thirsty were given to sucking dock-leaves? Can you imagine great Cæsar with a dock-leaf in his mouth?

Come here in April when dock grows at the bottom of this bank. Come down and see the leaves. No matter if you do tumble. It is as well to tumble down for knowledge sometimes as to climb for it. All knowledge is not on the heights.

Here on the back of a dock-leaf are the Brookside Folk that I would have you see. Little black mites, but destined to be beautiful greenwinged small beetles, — representatives of the *Chrysomelidæ*. Many a time you may notice the marred dock-leaves by the roads as you pass, or catch a glimpse of the small, yellow clusters of eggs of these beetles on some stem, or turned-over leaf. The larvæ have such a habit of dropping from the leaves that I wonder the motherbeetles place their eggs so near this brook. I should think that the children would fall in. Probably many a miserable little black mite meets this sad fate.

Keep some of these larvæ in a bottle with some soft, damp earth at the bottom, and with fresh dock-leaves, and you will find that when full grown the larvæ will descend and make for themselves little burrows in the soil where they will be transformed to light yellow pupæ. Put one of these pupæ under a microscope, and you will see the short, tiny, black hairs reaching out from the head and the dark dots standing for eyes. Does he know, as he lies motionless in yellow slumber, that he is being turned about and looked at? He will know what is done to him shortly, for you shall some **day go** to his bottle and find him waiting for you with life and motion in him and a perhaps **dull** green **vesture on.** If **you** come **early** enough **you** will find him **a** yellow beetle, with some iridescent shades about him perhaps. He is taking his first feeble footsteps and is somewhat inclined to tumble **over.** I left one such person as a yellow **beetle** one evening, and woke up the next morning to find him a green one; not so brilliantly green as are specimens that one often meets, but a dull color, still decidedly green.

In that hole **in the root** of that live-oak by **the** little bridge across the creek reside Mr. and Mrs. Sow-bug with their progeny. **In** that hole, alas! I once, digging with zeal **and a** trowel, caught a glimpse of a tail rapidly disappearing in the débris. My efforts were in vain. I could not catch **up.** What was **he?** Salamander? Who shall **tell?** Such glimpses are reminders that there are secret apartments in some of these trees where hermits may live, and they no doubt are rightfully indignant when a trowel reaches in and disturbs their meditations.

Some of these hermits live so far in that they cannot be reached till the tree is down. Such **are** some of the fat, **white larvæ** of beetles that **inhabit** live-oak trees. An **oak of** this kind was **chopped** down about a mile from here, and a friend of mine coming upon the spot gathered some of the numerous beetle-larvæ that sprinkled the ground.

Two that were brought me were great, fat white creatures. One when stretched out was about three and a quarter inches long and half an inch broad; the other measured two and a half inches in length.

The trunk of the tree, according to my friend's account, was riddled with the holes of the larvæ, and the report was that when the trunk first fell it was swarming with them. Those that I had made no more holes, however, seeming to be content to burrow in the earth at the bottom of their jar, and nibble the live-oak wood put in for their benefit. Six little prickles had each larva in lieu of legs, and small reddish spiracles marked the segments of the bodies.

Oak-tree larva of Beetle.

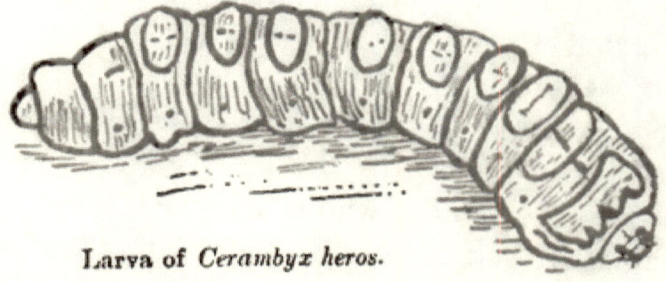

Larva of *Cerambyx heros*.

It must be very discouraging to be knocked

out of a hole that one has made for one's self with the intention of abiding in it till the larva-life is done. **Whether it** was this discouragement or not, from **some cause** my larvæ pined, grew flabby in flesh, and evidently gave up their prospects of becoming beetles. After living with me a few weeks **a larva died.** About five months after coming into **my** possession, a second larva departed from life. **He had shri**velled till he was but a faint image of **his** former plump self. Evidently it is necessary that such creatures should **be** allowed to **be** hermits **and** dwell in the interior of oak-trees till perfection **is** reached.

I think if my larvæ had lived they would have turned **to** representatives **of the** *Cerambycidæ*, the *Longicornia* of Latreille, the "Fiddlers" of the Germans, since **my** larvæ **had the** general shape of the children **of** that family, being larger in front than behind, having six little prickles to represent feet, and having the rings of the body furnished with humps. These were probably useful in dragging the **larvæ** through their holes. De Mouffet, **the ancient** and credulous, has this to say: "**Terambus,** a satirist, did not abstain from quipping **of** the **Muses,** whereupon they transformed him into a Beetle called *Cerambyx*, and that deservedly, to endure a double punishment, for he hath legs weak that he goes lame, and like a thief he hangs on a **tree.**"

The long, recurved antennæ of some of these

beetles led to a strange story that De Mouffet did his best to perpetuate by writing it down. In speaking of one kind of the family, the Prionus, and mentioning the antennæ under the name of "two horns that grow above their eyes," he says, "they are flexible with nine or ten joynts; not exactly round, but are rough like goats' horns, which, although it can move them every way, yet when it flies it holds them only forth directly; and being wearied with flying, she useth them for feet; for knowing that his legs are weak, he twists his horns about the branch of a tree, and so he hangs at ease, as our Bruerus saw in the country about Heidelberg; in that it resembles the Bird of Paradise which, wanting feet, clings about the boughs with those pendulous nerves, and so, being tired with labors, takes its ease."

I do not know whether the old naturalist always meant antennæ by "horns." I think he sometimes meant jaws, for he says of one beetle, it "useth its hornes for that end for which crabs and lobsters do their clawes." But he says, "Beetles are some greater, some less. The great ones, some have horns, others are without horns. Those that have horns, some are like Hart's horns, others like Goat's horns, . . . others have ram's horns; some have horns on their nose."

I have read that an English entomologist named Drury has recommended to all those who are ever cast upon desolate islands where nothing can be found to eat, that they should search the tree-trunks for such grubs as feed on wood; and he says that people for a short time, at least, can comfortably live on such grubs.

Certainly one would have to be somewhat hungry to partake of such diet. I imagine, however, that all the courage would be needed for the first bite. It would almost be easier to partake of the

Live-oak Cocoons. Moths. Cocoons about life-size.

painted paper-snakes wherewith Agnolo della Pergola fed his prisoner, Zanobi del Pino, telling him that of a Guelph they would make him a Ghibelline.

Other things, found on the live-oak leaves, are numbers of small, white cocoons. I have found them with what looked like small threads running their length, and in November some have a round hole in one end. Gather some of the cocoon-dotted oak-leaves in the spring. They are easily found, after your eye is once used to looking for them, and in May you will find in your paper-covered bottle small, light-colored moths, looking much like clothes-moths, but smaller if anything, although had I seen them fluttering in the house I should have pronounced them those Tineans of which all good women have such a dread. But look at the oak-moths closely and you will see that their wings are marked with yellow.

In April on that old willow by the fence where the path is just wide enough for one to walk without tumbling off, I found a lady-bug. Not the common kind, but the smaller black variety with a red spot on either wing. *Chilocorus* is my lady's genus, I believe. The one I found was either of the same variety as the "twice-stabbed" lady-bug that has rendered such service in warring against the scale-insects in California orchards, or was next cousin to that useful beetle. Who was it that first suggested that one or two mashed lady-bugs put into the hollow of an aching tooth would stop the pain? I read of a naturalist who attempted curing his tooth-ache with this remedy, and he professed to have been relieved. Little he deserved it, though, according to my ideas. What! Shall a mortal take the *Marien-Käfer*, the "Lady-beetles of the Virgin," the *Vaches de Dieu*, and smash them to cure a vile tooth-ache! "To what base uses we may return, Horatio!"

Of course where plant-lice are, there come the larvæ of the common lady-bugs. Rather ferocious-looking, but **very** harmless creatures they are. A person will always know the eggs **after** once seeing them. **They are** bright yellow, **oval,** standing perpendicularly on **one** end **in** a manner that Columbus' **egg might** have envied.

Larva of Lady-bug. *Hippodamia convergens.*

There were seven eggs that **I** found once on the back **of a red** rose-leaf, but **the** eggs turned **gray** just **before** hatching, and something of the **creatures** within could be seen through **the** walls.

I saw several larvæ just coming **out** of those **eggs.** A hole would appear **in** the top of an egg **and the larva's head would** appear. Then the **creature** would slowly come out, pulling first one foot free and then another, till all six were **loose from the egg.** Two larvæ **lay on** their backs, going through **this operation. I finally** helped **one** free and put him on a leaf where he lay exhausted. **The legs of these** larvæ look preposterously long **when they first come** from the **egg and** have not yet walked. After **a** number of hours the bodies **of the larvæ** become darker, and the looker-on can **recognize the** regular lady-bug-grub look. I have raised numbers **of** thirteen-spot lady-bugs from such larvæ.

There are a great many Aphides beside this **brook.** One finds the creatures on very varying kinds of plants, and has occasionally to brush the crawling things from one's cape. But the enemies of Aphides are abroad, too. The Syrphus flies hover around one's pathway, looking like little wasps. The *Syrphidæ* will see to it that plenty of their larvæ go forth to slay the Aphides.

The grub have a way of holding to the leaves with the hind end of the body and stretching out the extensile forward portion, waving it about after the manner of a blind man feeling with his stick. For these Syrphus grubs are blind and footless. The wasp-like flies place their eggs among groups of Aphides, usually not more than one or two eggs on a leaf. The eggs I have found have been white, oval, and about one thirty-second of an inch long. When the grub comes forth it has not far to reach to obtain its food, a convenient arrangement for blind folk, although I did once find an egg on a rose-twig where not a single aphis could be seen, even under the microscope. I suppose there are giddy individuals even among Syrphus flies, but that mother certainly deserved a reprimand for placing her poor blind infant in such a situation that it would have to travel all over the branch, hunting for its first meal. On inspection, however, I found something on the shoot that may have been the white skeleton of a dead aphis. Perhaps that aphis was there alive and well when mother Syr-

Syrphus grub eating Aphis.

phus deposited that egg. Perhaps she thought that one aphis was enough for her infant's first meal. She did not want him to injure himself by over-eating. Let us be charitable.

You can see such grubs pick up the Aphides one by one. How must it seem to reach out, catch one's breakfast, hold it up in the air, even if it is as big as one's self, and devour it alive?

I have had Syrphus larvæ that had a tinge of red in their complexions. When a grub has eaten enough, it sticks itself to a leaf, its body draws up and becomes somewhat hard. After a time this puparium opens and the fly appears. One that I raised had a golden thorax and a gold-and-black abdomen, the legs being yellowish, and the eyes reddish and big. The body was slightly tinged with green underneath.

Great numbers of these flies must appear in the course of a year. I have found grubs at work in April, and I know not how long before such creatures may have begun to stir. I have found other grubs that came out as flies as late as October. We owe much to these diligent blind grubs. If a person can put enough larvæ of the Golden-eyed Fly, Syrphus, and the Lady-bug on trees infested with plant-lice, there need be little fear that the enemy will not be conquered. There are many little servants ready to aid man if he will but become acquainted with them and show them the work he wants them to do. "All flies shall perish except one, and that is the bee-fly," says the Koran. I know not whether the writer of that sentence meant Syrphus. I doubt it, but surely that useful insect deserves to live longer than many others. Not for such as it should be such a " Papal Cursing Bell" as that which Horace Walpole on Strawberry Hill kept for a curiosity, the bell being that made by Benvenuto Cellini for Pope Clement VII., and being formed of silver, carved outside to represent serpents, flies, grasshoppers, and other baleful creatures that were to be warded from the lands of the faithful. I trust that the old Popes were versed in entomology, and never banned the *Syrphidæ* or like useful creatures.

If the ecclesiastics trusted entirely to the observations of country folk, however, in discriminating between insect foes and friends, I much fear that mistakes were sometimes made, or else the people of those days were brighter than persons now. For it sometimes takes a deal of looking to know friends from foes, especially if the looker has no books to help him in the search. And yet a friend of mine once told me that she had been in farm-houses in the country where she saw Reports of the Department of Agriculture, containing information about insects that farmers ought to be acquainted with, sold for old paper! Surely such people deserve all the insect enemies that come to see them.

"We that have good wits have much to answer for," says Touchstone, but, if that plain-spoken clown would ever come out of that forest of Arden in which he perpetually lingers, perhaps he might acknowledge that those who have *not* good wits in this day have much to answer for, and are obliged to make answer, too, when they see their crops devoured by creatures that they might have known how to fight, if they had read the information freely placed in their hands. And perhaps Touchstone, beholding the multitude of the insect creation, might again have occasion to remark, "The fool doth think that he is wise, but the wise man knows himself to be a fool," — a saying that quite expresses the ideas of people in regard to " bugs;" those who know but little about them being much more elated with self-wisdom than those who know more.

I once had the privilege of giving a scarcely one-day-old, perhaps not more than half-a-day-old, Syrphus larva his first meal. He was a brave youth and had started out independently enough to seek his own fortune, when I, like a fairy godmother gave him my gift. It was an aphis about his size. He stuck his head into the aphis, back of the victim's head and went to work.

For forty-five long minutes I watched him. He sucked the aphis as a baby would a bottle. Tiring of his devotion to that aphis, I took it from him. Poor fellow! It was his first disappointment in life. He fell prone under it at first, and then began to wildly stretch around demanding his prize again. I put him on a leaf and allowed him to search for an aphis to his liking. But he was hard to suit. I think he had eaten enough, for he walked among the Aphides awhile, an uneasy speck, threatening them with destruction, yet harming none of them. So I thrust him out into the world and let him go on his blind way to shift for himself in the branches of a pink moss-rose. Probably before now he has brought ruin and devastation to many a happy aphis. It is wonderful how independent insects are, even from the first moment they gain strength after coming out from the egg. Without a mother to care for them or a home to shelter them, out they go. The whole world is their home and they are happy in it. Let enemies come and they are ready for them, sometimes more than ready, eager for the conflict, having the spirit of the Templars, of whom it is said that none ever asked the number of the enemy, but only demanded, "Where are they?"

But how hard-hearted are the insects toward one another! In all the time that I have watched them I do not recollect ever having seen an act of compassion performed by any kind of insect for another. The brown plant-lice and the ants that I have found together on the backs of willow-leaves seem to be perhaps as friendly as any insects that I remember at present, and yet I doubt if their relations are prompted by kindness.

Do not think that all is joy beside this brook. To those who have eyes to read, there are records here of many blighted hopes. A fence runs beside this brook, and on it, as on others around here, you may find cocoons in their season. But poke your stick into one, and you will find, not a healthy-looking pupa of a caterpillar that shall come out as a moth, but a number of little brown pupæ-cases, much smaller than the caterpillar. The cocoon may also contain a remnant of the caterpillar's dried up body.

Many a caterpillar has crawled up these boards and formed for himself a cocoon in some sheltered nook, and, even while forming it, he must have been aware of certain very queer and painful feelings inside of himself. Awful things went on inside that cocoon, things never revealed save to those persons who go out some day to gather cocoons.

If you wish to have a crop of flies in the spring, shake the October pupæ-cases from the cocoons into your bottle, put it away covered with mosquito-bar. Some March day you will look into the bottle and will see four or five flies longing for freedom. They are about as large as houseflies, and these villainous beings have lived as maggots inside of the caterpillars and have prevented some moths from ever appearing in the world. I have found several such maggots in a single cocoon. In raising caterpillars, one is almost sure to have some from whose bodies will come maggots. Give such fly-worms a bit of earth, and in an incredibly short time they will transform to brown pupæ.

Other inhabitants of the cocoons are the *Dermestidæ*, or Skin-beetles, and their children. Going up that hill one day early in September, when all the world was in dust, and when the evil-smelling daisies were almost the only flowers in bloom, I saw a cocoon on the fence. Poking into the cocoon I found a perfect beetle and five brown and white larva of the *Dermestidæ*. The larvæ wore trailing hairs, like a lady's train, behind them, and had a bunch of hair on each side of each segment.

As for their mother (for I have no doubt she stood in that relation to them), her brown back was marked with white, and she was a deceitful person to have the training of any children; for if she was disturbed, down she tumbled on her back and pretended to be dead. The infants did not seem to care how many times she died. Probably they were used to her deaths.

Larva of one of the *Dermestidæ*, — enlarged.

After keeping them a time I thought that the infants were also dead. They lay motionless in the bottom of the bottle, and bitterly did I accuse myself of having been their murderer, by not having given them caterpillar or cocoon remnants enough for their sustenance. But hypocrisy ran in the family. In October I tipped out the contents of the bottle. There was the remnant of a chrysalis, some skins of larvæ, and some caterpillar hair. There were also two or three beetles. The larvæ had not been dead at all.

The new beetles were like their parent in body and mind, for one immediately fell on his back, his feet folded, and every feature exactly like death. I helped one of the beetles out of the remnant of a pupa case that still stuck to his back. But he was very dead during the operation.

I remember sitting mourning over a beetle of this kind once. I held him in my hand, and as I wondered why he died, he moved a leg! I have never become accustomed to the perfection to which the skin-beetles carry their mimicry, and if I have to decide a case of death I usually gaze at the beetle as long as my patience holds out, and if he has not moved then, I leave him, knowing no more than I did before inspection. Unless such a beetle is positively shrivelled, no one can tell whether he is dead or not.

And so the traveller beside this brook and over these hills may learn, if he looks, that man is not the only creature who builds houses and is disappointed about living in them. There is material here for a fine sermon, after all, take the brook through and through. Here are fightings and murders and thefts and trickeries, the semblance of death, the awakening from slumber, the rising to new life, the change from the grovelling on the earth to the soaring of wings in the sunshine.

Indeed, I doubt if there is any corner of the world where one may not find one's text and preach a sermon to one's self, if in a sermonizing mood. For myself, there is in these brooks something as eloquent of Him whom they of old called the "All-Father," as there might be for me in any other nook of the universe. His hand has been here also. Let us conclude with the words that old De Mouffet wrote two hundred and fifty years ago, for they are as true now as then, "*All things are full of God's wonderfulnesse.*"

INDEX OF ILLUSTRATIONS.

Aphis, Syrphus grub eating, 212.
Architecture, A triumph of, 142.

Beetle, Oak-tree larva of, 206.
Belostoma grandis, 22.
Blackberry leaf, Eggs on back of, 133.
Blue-bells, 80.
Brodiæa terrestris, Bud and flower of, 80.
Bug, Little, a day old, 133.
Butterfly Lily, 82.

Caddis-worm's House with "logs," The bigger, 140.
California Hydrophilidæ, 30.
Calochortus Weedii, 82.
Case with Caddis-fly larva, 135.
Catch-poll, 9.
Caterpillar of Orgyia, 194.
Cerambyx heros, Larva of, 206.
"Chore," Sandy's unfinished, 138.
Chrysopa, 202.
Chrysopa, Cocoons of, 200.
Chrysopa, Eggs of, 196.
Chrysopa larva, 198.
Cocoon of Orgyia, 195.
Cocoons of Chrysopa, 200.
Communis, Cyclops, 122.
Corydalus cornutus, Pupa of, 153.
Corydalus, Horned, 151.
Corydalus, Larva of Horned, 147.
Cyclops communis, 122.
Cypris unifasciata, 122.

Dermestidæ, Larva of, 218.
Dodecatheon Meadia, 81.

Dragon-fly larva, Large, 6.
Dredger, Home-made, 75.
Dytiscus, 3.
Dytiscus marginalis, Larva of, 59.
Dytiscus pupæ, 73.

Eggs of Chrysopa, 196.
Eggs of Hydrophilidæ, 35.
Eggs of Water-boatman, 55.
Eggs on back of blackberry leaf, 133.
Epaulette, Scorpion-bug bearing, 18.

Fly, Golden-eyed, 202.
Fly, Tipulid, 88.
Frog-hopper, Larvæ of, 105.

Golden-eyed Fly, The, 202.
Gordius aquaticus, 118.

Hippodamia convergens, 211.
Home-made Dredger, 75.
Horned Corydalus, 151.
Hydrachna geographica, 124.
Hydrophilidæ, California, 30.
Hydrophilidæ, Eggs of, 35.
Hydrophilidæ, Larvæ of, 36.
Hydrotrechus remigis, 2.

Lady-bug, Larva of, 211.
Larvæ of Hydrophilidæ, 36.
Larvæ, Red Dragon-fly, 48.
Larva, Case with Caddis-fly, 135.
Larva, Chrysopa, 198.
Larva, Large Dragon-fly, 6.
Larva, Mask of Libellula Dragon-fly, 9.
Larva, Water-mite, 124.
Larva of Cerambyx heros, 206.

INDEX OF ILLUSTRATIONS.

Larva of Dermestidæ, 218.
Larva of Dytiscus marginalis, 59.
Larva of Frog-hopper, 105.
Larva of Horned Corydalus, 147.
Larva of Lady-bug, 211.
Leaf I found, The, 196.
Libellula Dragon-fly **larva**, Mask of, 9.
Lily, Butterfly, 82.
Lily, Mariposa, 82.
Live-oak cocoons, 209.

Mariposa Lily, 82.
Mask of Libellula **Dragon-fly** larva, 9.
Mimulus **glutinosus**, 81.
Monarch at rest, The, 73.
Moths, 209.
Moth, Winged male, 195.
Moth, Wingless female, 195.
Mourning Cloak, 181.
My Ranatra, 23.

Notonecta glauca, 52.
Notonectidæ, 4.

Oak-tree larva of Beetle, 206.
Orgyia, Caterpillar of, 194.
Orgyia, Cocoon of, 195.
Orgyia Leucostigma, 195.

Papilio Turnus, 140.
Pupæ of Tipulidæ, 87.
Pupa of Corydalus **cornutus,** 153.

Ranatra, My, 23.
Red Dragon-fly larvæ, 48.

Saudy's unfinished "**chore**," 138.
Scorpion-bug bearing epaulette, 18.
Shooting-star, 81.
Syrphus grub eating Aphis, 212.

Thysanura, One of the, 173.
Tipulidæ, Pupæ of, 87.
Tipulid Fly, 88.
Triton, 104.
Triumph of Architecture, A, 142.

Vanessa Antiopa, 180.

Water-boatman, Eggs **of,** 55.
Water-boatmen, 4.
Water-mite Adult, 124.
Water-mite Larva, 120.
Water-scorpion, 17.
Water-skater, 2.
Winged male Moth, 195.
Wingless female Moth, 195.

The Riverside Library
for Young People.

A Series of Volumes devoted to History, Biography, Mechanics, Travel, Natural History, and Adventure. With Maps, Portraits, etc., where needed for fuller illustration of the volume. Each, uniform, strongly bound in cloth, 16mo, 200–250 pages, 75 cents.

1. *The War of Independence.*
   By JOHN FISKE. With Maps.
2. *George Washington: An Historical Biography.*
   By HORACE E. SCUDDER. With Portrait and Illustrations.
3. *Birds through an Opera Glass.*
   By FLORENCE A. MERRIAM. Illustrated.
4. *Up and Down the Brooks.*
   By MARY E. BAMFORD. Illustrated.
5. *Coal and the Coal Mines.*
   By HOMER GREENE. Illustrated.
6. *A New England Girlhood, Outlined from Memory.*
   By LUCY LARCOM.
7. *Java: The Pearl of the East.*
   By Mrs. S. J. HIGGINSON. With a Map.
8. *Girls and Women.*
   By E. CHESTER.
9. *A Book of Famous Verse.*
   Selected and arranged by AGNES REPPLIER.
10. *Japan: In History, Folk-Lore, and Art.*
    By WILLIAM ELLIOT GRIFFIS, D. D.
11. *Brave Little Holland, and What she Taught us.*
    By WILLIAM ELLIOT GRIFFIS, D. D.

(Others in Preparation.)

MESSRS. HOUGHTON, MIFFLIN AND COMPANY publish, under the above title, a series of books designed especially for boys and girls who are laying the foundation of private libraries. The books in this series are not ephemeral publications, to be read hastily and quickly forgotten; both the authors and the subjects treated indicate that they are books to last.

The great subjects of History, Biography, Mechanics, Travel, Natural History, Adventure, and kindred themes form the principal portion of the library. The authors engaged are for the most part writers who already have won attention, but the publishers give a hospitable reception to all who may have something worth saying to the young, and the power to say it in good English and in an attractive manner. The books in this Library are intended particularly for young people, but they will not be written in what has been well called the *Childese* dialect.

The books are illustrated whenever the subject treated needs illustration; history and travel are accompanied by maps; history and biography by portraits; but the aim is to make the accompaniments to the text real additions.

The publishers hope to have the active coöperation of parents, teachers, superintendents, and all who are interested in the formation of good taste in reading among young people.

HOUGHTON, MIFFLIN AND COMPANY,
4 Park Street, Boston; 11 East 17th Street, New York.

www.ingramcontent.com/pod-product-compliance
Lightning Source LLC
Chambersburg PA
CBHW021821230426
43669CB00008B/821